SpringerBriefs in Electrical and Computer Engineering

Signal Processing

Series editors

Woon-Seng Gan, Singapore, Singapore
C.-C. Jay Kuo, Los Angeles, USA
Thomas Fang Zheng, Beijing, China
Mauro Barni, Siena, Italy

T0275980

More information about this series at http://www.springer.com/series/11560

Patrick Bas · Teddy Furon
François Cayre · Gwenaël Doërr
Benjamin Mathon

Watermarking Security

Fundamentals, Secure Designs and Attacks

 Springer

Patrick Bas
Centre National de la Recherche Scientifique
  (CNRS)
Lille Cedex
France

Teddy Furon
INRIA Rennes
Rennes
France

François Cayre
Grenoble Institute of Technology
Grenoble
France

Gwenaël Doërr
Technicolor R&D France
Cesson-Sévigné
France

Benjamin Mathon
Université de Lille
Villeneuve-d'Ascq Cedex
France

ISSN 2191-8112   ISSN 2191-8120   (electronic)
SpringerBriefs in Electrical and Computer Engineering
ISSN 2196-4076   ISSN 2196-4084   (electronic)
SpringerBriefs in Signal Processing
ISBN 978-981-10-0505-3   ISBN 978-981-10-0506-0   (eBook)
DOI 10.1007/978-981-10-0506-0

Library of Congress Control Number: 2016942495

Printed on acid-free paper

This Springer imprint is published by Springer Nature
The registered company is Springer Science+Business Media Singapore Pte Ltd.

Foreword

The interest in watermarking surged in the second half of the 1990s when watermarking was seen as a fundamental ingredient of the Digital Right Management systems being developed in that period, as a possible solution to protect multimedia assets, noticeably digital video and audio, from copyright infringements. As such, it was clear since the very beginning that watermarking techniques had to operate in highly hostile scenarios characterized by the presence of one or more adversaries aiming at deleting the watermark from the host document or at least to make it unreadable. Yet it took several years before researchers would start casting watermarking security studies into a correct framework. For almost a decade robustness and security were considered interchangeable terms. In the most favourable cases, security was merely seen as a higher level of robustness. At the beginning of 2000s watermarking researchers started looking for a precise definition of watermarking security. Such an effort led to a deep understanding of the difference between watermarking robustness and security. In a few years a bulk of theoretical and practical studies emerged establishing watermarking security as a flourishing and well-defined research stream within the wide body of watermarking scientific" literature. Unfortunately, very few attempts have been made to gather at least the main results concerning watermarking security into a unique publication, so that it is still common for newcomers (and not only) to treat watermarking security in a naïf way, and confuse security and robustness as in the early 1990s.

In the above framework, this Springer Brief is a welcome effort, which will help researchers and practitioners to get a comprehensive view of watermarking security as it has been developed in the last decade. Being one of the researchers who most contributed to shape watermarking security as a theoretically sound discipline and to turn the theoretical findings into practical solutions, Patrick Bas along with his collaborators has leveraged on their personal experiences to write an excellent book that guides the readers to understand even the most difficult concepts and to show how to exploit the theoretical results to design a secure watermarking system which is on par with the state of the art in the field.

The book starts with an introductory chapter in which the author motivates the need to study watermarking security by the light of the most recent developments. Then, Chap. 2 introduces the basic concepts of watermarking theory. While experienced readers can skip this part, Chap. 2 represents a successful example of a synthetic overview of the main concepts of watermarking theory, which is worth reading. Chap. 3 introduces the main concepts underlying watermarking security from a theoretical point of view, while Chap. 4 illustrates how the theoretical findings can be applied to design a secure watermarking system in practice. The last part of the book is devoted to the description of some of the most serious attacks against watermarking security. It is well known, in fact, that the real security of a system can be understood only by trying to attack it. Watermarking is no exception, thus justifying the attention paid within the whole book to describe the possible actions of the attacker(s).

The entire book is pervaded by a praiseworthy pedagogical effort aiming at making accessible even the toughest concepts. The result is an enjoyable and easy to read book, which I am sure the readers will appreciate as much as I did.

Mauro Barni
University of Siena

Contents

Chapter 1
General Introduction

1.1 Forewords

This book couldn't have been written without a collaborative work between the main author (Patrick Bas) and his colleagues.

In Chap. 3, the work on watermarking security classes comes from a collaboration with François Cayre, and the work on the key length of watermarking scheme comes from a joint work with Teddy Furon.

In Chap. 4, the seminal ideas of secure spread-spectrum watermarking come from a collaboration with François Cayre, and the idea of using optimal transport for spread spectrum scheme has been found with Benjamin Mathon and François Cayre. The Broken Arrows scheme is the result of a joint work with Teddy Furon during the development of the BOWS2 challenge.

In Chap. 5, the work on the security analysis of Trellis based watermarking scheme is a joint work with Gwenaël Doërr.

1.2 Notations

- As a general rule and without other mentions, a represents a real value, A a random variable associated to a, $\mathbf{a}$ a vector represented as a column vector, $\mathbf{A}$ a matrix, and A a random vector. The roman font, for example a(.) is reserved for functions, symbols or names. $\mathbf{a}^T$ and $\mathbf{A}^T$ are respectively the transposes of column vector $\mathbf{a}$ (hence a row vector) and matrix $\mathbf{A}$. $\mathbf{a}[i]$ represent the ith element of vector $\mathbf{a}$.
- $\mathbf{c} \in \mathcal{C}$: the host content
- $\mathbf{x} \in \mathcal{X} = \mathbb{R}^{N_v}$: the content vector to be watermarked, extracted from $\mathbf{c}$, $\mathcal{X}$ is called the embedding space
- $\mathbf{y} \in \mathcal{X} = \mathbb{R}^{N_v}$: the watermarked vector
- $\mathbf{w} = \mathbf{y} - \mathbf{x} \in \mathcal{X} = \mathbb{R}^{N_v}$: the watermark vector
- $\mathbf{z} \in \mathbb{R}^{N_v}$: the watermarked vector corrupted by noise

© The Author(s) 2016
P. Bas et al., *Watermarking Security*, SpringerBriefs in Signal Processing,
DOI 10.1007/978-981-10-0506-0_1

- $\mathbf{n} \in \mathbb{R}^{N_v}$: the noise added to the watermarked content
- N_v: dimension of embedding space
- $\mathbf{k} \in \mathcal{K} = \mathbb{R}^{N_k}$: a vector representing the secret key, $\mathcal{K}$ is called the key space.
- N_k: the dimension of the secret key
- N_r: number of secret regions used during the embedding
- $p_A(\mathbf{a})$: the probability density function of the random vector $\mathbf{A}$
- $P_A(a)$: the probability density function of the random variable A
- N_m: the size of the embedded alphabet
- N_b: the number of embedded bits ($N_m = 2^{N_b}$)
- $D = E[||\mathbf{w}||^2]$: the embedding distortion
- m: the embedded symbol, $m \in \{1, \ldots, N_m\}$
- N_c, the number of codewords coding the same symbol m
- N_o, is the number of observed contents
- $u(.)$: the Heaviside function
- $|s|$: the absolute value of s
- $||\mathbf{v}||$: the L-2 norm of $\mathbf{v}$
- AWGN: the Additive White Gaussian Noise
- $\mathbb{P}(.)$ the probability of an event
- p_{md}: the probability of miss detection, probability to not detect a watermark on a watermarked content
- p_{fd}: the probability of false detection, probability to detect a watermark in an original content

1.3   Watermarking and Security

Because the goal of each technology is an answer to a set of pre-defined requirements, it is often very complicated to write "secure technology" inside this list of requirements. This is due to the fact that security can hardly be defined as a set of known constraints since it consists in adding an actor whose goal is to divert the technology from its initial meaning. Consequently, taking into account security constraints is a complicated task since it relies on the power of the adversary but also on his goals. As an example, it is possible to specify the construction of a bridge in order to convey different categories of vehicles, or to cope with different classes of earthquakes, but how to design it in order to cope with an unknown attack? Coping with security consequently implies the virtual presence of an adversary inside the development process, but also the definition of flexible requirements in order to cope with a large range of adversaries.

In order to address the problem of watermarking from a rational point of view, we can define watermarking as the embedding of a **robust, imperceptible and secure** information. The robustness of the system allows us to decode or detect the watermark if the content undergoes different usual transforms, its imperceptibility guaranties that its artistic or pay value does not decrease once watermarked, and the security

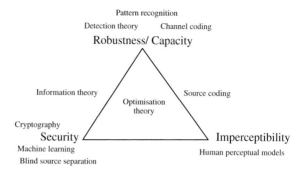

Fig. 1.1 Fundamental constraints in watermarking and its associated academic fields

guaranties that the embedded information is safe whenever the system has to cope with an adversary.

In watermarking the constraints are numerous: contrary to image denoising for example, our goal is not to optimise a constraint such as the signal to noise ratio but to optimise one of the three described above, while not lowering one of the others. Moreover, each of these constraints can be refined according to the targeted application and is associated with various research fields. The list of related topics is not exhaustive but we can nevertheless refer to digital communications (especially source coding and channel coding), information theory, detection theory, optimisation theory, pattern recognition, psychovisual models, cryptography, blind source separation or machine learning. As illustrated on Fig. 1.1, these scientific fields can be seen inside the triangle of constraints specific to watermarking, they can be either specific to one of the three constraints, or associated to several of them. We also have to remember that for a targeted application, the constraints related to implementation are of prime importance since they allow for example real time embedding or decoding on different bitstreams such as images, audio or video.

In this book we deal with the problem of watermarking by focusing on security constraints and we provide a theoretical and practical framework to this problem. We present different ways to increase the security of a watermarking scheme or on the contrary we incarnate the adversary by designing attacks trying to erase, modify or copy the embedded message. We also study what is the impact of security on robustness and imperceptibility: are theses constraints compatible or antagonist?

1.3.1 The Needs for Watermarking Security

Our society is now digital: in this digital word people share their ideas, their feelings, their data in a completely natural and friendly way. The tools that enable these interactions are for example the blogs, the RSS feeds, the forums, or the sharing servers located on the cloud. In the digital world, the smallest technical or personal problem can be shared and even solved in few seconds by looking at the distributed knowledge or by requesting the help of the connected people. In the digital world, a user has access to a nearly infinite amount of information, and mostly for free.

The counterpart of this freedom is that for a lot of people, the notion of copyrights associated to digital contents has become more and more vague: if the act of stealing a music tape during the 80s required a certain amount of energy, it only requires a USB key or an Internet connection in the digital world to get a music album and this in few seconds. In order to fight against this digital leakage, the multimedia industry has tried since the end of the last century to develop technologies that enable copyright protection. The watermarking technologies were born at this time in order to provide solutions in this context.

For illustrative purposes, we describe 4 examples of watermarking usage, either for copyright protection (example 1, 2 and 3) or to increase the functionality of a multimedia content (example 4):

1. After failing to protect DVD disks, the Blu-ray format had to come up with better protection systems. Its associated standard, called BD+, gives the possibility to the editor to use two different watermarking technologies to protect their contents. The first one is based on the audio watermarking technology called Cinavia[1] [1] and its goal is to forbid displaying movies that were previously ripped (deciphered). If a watermark is detected on a clear Blu-ray, this means that the content is a copy of a ripped disk and the player consequently prevents the rendering.

   The second anti-copy protection technique embeds a watermark while playing the content in order to identify the Blu-ray player [2, 3]. The watermark is embedded on the fly using a clever and simple process: a subset of the video exists in two different but perceptibly identical versions, each containing its own watermark (0 or 1), on the disk. Embedding is performed by playing one version or the other according to its identification number (called sequence key). If the video is found on an illegal network, the Blu-ray player can be identified and then black-listed: it will not be allowed to play the future Blu-ray releases.

2. Because theatres are also a potential leakage source toward illegal sharing networks, the watermarking industry offers solutions to identify theatres where movies are projected then captured with a camcorder, the films are then watermarked with an ID specific to each theatre. Theses techniques [4] can be used to first find the location of the theatres and thereafter to stop new attempts of recording on site.

3. One of the most popular use of watermarking was to identify a leaker (this principle is called "traitor tracing") during the Oscar voting process in 2004.[2] This person has put his personal copy on a sharing network and the watermark associated to the copy was used to trace back the leaker.

4. It is also worth noticing that watermarking can also be used in other scenario than copyright protection. For example the use of a "Second Screen"[3] uses a watermark

---

[1]www.cinavia.com.

[2]http://en.wikipedia.org/wiki/Carmine_Caridi.

[3]http://www.fastcompany.com/1722925/greys-anatomy-goes-dual-screen-thursday-with-this-sync-ipad-app.

embedded in audio files of TV series in order to give extra information on the content, which is read by another device such as a mobile phone or a tablet[4] in order to increase the ambient information.

1.3.2   Who Are the Adversaries?

In order to guess his goal and his power, it is important but not easy to have an accurate picture of the adversary. We propose here a typology inspired from [5, 6] which presents different types of *hackers*, i.e. persons whose goal is to bypass a technology.

1. The *mafias* and the *cyber thieves* whose goals are to find weaknesses of information systems in order to make money out of them. They look for example for information access to bank accounts as the thief in 2011 of data saved by the Sony Playstation system.[5]
2. The *information miners*, whose goals are to get or to generate information. The generation process will be performed for example by the *spammers* or *phishers* who send emails in order to trap internet users and get their credentials. They are often working for *Mafias*. Information miners also try to get into information systems in order to perform industrial espionage. We can also add in this category people wanting to bypass private information in order to do public defamation.
3. The *ideologues* who want to harm their adversaries by hacking their information systems. Examples are "hacktivist" groups such as Anonymous or Lulzsec who look to unveil pitfalls of information systems that collect private information,[6] attacks on bank websites which are enemies of Wikileaks[7] or attacks against enemies of Wikileaks to shut down their website.[8]
4. The *digital soldiers* who try to harm military organisations or states by destructing their information systems as it was done on Iranian nuclear plants using the Stuxnet virus.[9]
5. The *technophiles* who want to bypass a protection system in order to increase the features of their device (a smartphone for example), they are often thrilled by performing technological exploits.

Even if the power and the dangerousness of the adversaries always rely on their technical skills and their ability to work in network, we can try to quantify the specific power of each adversary.

---

[4]https://www.digimarc.com/discover/.

[5]http://www.wired.com/threatlevel/2011/04/playstation_hack/.

[6]http://www.guardian.co.uk/technology/2011/jun/16/lulzsec-anonymous-hacking-games-companies.

[7]http://www.zdnet.com/news/wikileaks-hackers-attack-visa-get-banned-by-facebook-twitter/490442.

[8]http://www.zdnet.com.au/wikileaks-hit-with-ddos-attack-339307578.htm.

[9]http://fr.wikipedia.org/wiki/Stuxnet.

The national security agencies (who play the roles of digital soldiers or cyber thieves) can have a important power coming from both their skills and the important number of employees. As an example more than 30,000 people work for the american NSA today.

The mafias and the information miners are also organised and can also have an important power.

The power of the technophiles and the ideologues will rely on their skills (if many are not very skilled, some are extremely) and on their ability to gather in networks.

Coming back to watermaking, the adversaries who try to hack watermarking systems belong either to the *mafia* class (organisations or companies looking for a direct profit coming from their hacking software), or to the *technophiles* class that will try to hack the watermarking system for the benefit of a part of the population. Here are two examples that illustrate technophile hackers bypassing a protection system, the first example illustrates the possibly important power of this group, the second is directly linked to the use of a watermarking technology:

- the iPhone has been released the 29th of June, 2007, in a setup where the phone can only communicate using a specific operator. 19 days after the release, a software to bypass the protection was proposed by the "iPhone Dev Team" and allowed users to use any operator.[10] Since its release the device and its successors have passed through different security updates and different methods to bypass the updates.
- in watermarking, the Cinavia technology (see Sect. 1.3.1, 1) is also under the scrutiny of hackers. Several techniques have already been proposed to bypass the anti-copy protection system. In late 2015, two of theses methods can be considered as watermarking security attacks since they directly process the audio track of the movie to remove the watermark.

1.3.3 Standardising Watermarking Security

Standardisation committees have recently addressed the problem of risks management applied on Information Systems by creating the norm ISO/27005.[11] This norm tackles risk management by (1) measuring risk, (2) defining protection systems and (3) evaluating these systems.

In the context of watermarking, a similar approach has been developed in an independent way [7].

On important concept to invoke in order to deploy a security analysis of a watermarking scheme is the Kerckhoffs' principle.[12] This principle claims that the security of the watermarking scheme only relies on the use of a secret key. It assumes that the adversary is able, often using reverse engineering techniques, to understand and

---

[10]http://www.wired.com/gadgetlab/2007/10/a-brief-history/#more-6146.

[11]http://fr.wikipedia.org/wiki/ISO/CEI_27005.

[12]http://en.wikipedia.org/wiki/Kerckhoffs's_Principle.

to know the watermarking algorithm. Under the Kerckhoffs' principle, the goal of the adversary will be to estimate the *secret key*.

There are many possible definitions of a secret key, it can be for example a stream of characters that once converted into a number will seed a pseudo-random generator and will be used to generate pseudo random vectors. Here we prefer to use a more abstract definition: "a watermarking key" is tantamount to a set of parameters defining a set of detection or decoding regions $\mathcal{D} = \{\mathcal{D}_1, \ldots, \mathcal{D}_{N_r}\}$ in a space $\mathcal{X}$ of dimension N_v. Note here that the space $\mathcal{X}$ is a subspace of the space defining the host content **c**. When **c** is a digital image for example, **x** can denote a vector representing a set of pixels of the image, or a vector of coefficients generated from a known transform such as the Fourier transform or the wavelet transform, or a key dependent transform. For multi-bit watermarking the definition of the key can be extended to a bijective function linking each region $\mathcal{D}_i$ and a message $\mathbf{b}_i$ (see Sect. 2.1.4).

The watermark embedding consists in a displacement of the host vector $\mathbf{x} \in \mathcal{X}$ toward one of the region $\mathcal{D}_i$ of the set $\mathcal{D}$ in order to obtain the vector **y**. The goal of the adversary is to estimate the topology of the set of regions and possibly the mapping between the messages and the regions $\mathcal{D}$. If he succeeds in doing that, the adversary will be able, for a multi-bit scheme, to read the embedded message, to modify it or to copy it into another content. For a zero-bit watermarking scheme, the adversary will be able to detect the watermark or to modify the watermarked content while minimising the distortion in order to remove the watermark, or to copy the watermark into another content (see Fig. 1.2).

The material given to the adversary is also a mean to specify the security scenario. In this collection of papers [7–9], the authors propose different security scenarios, each of them associated to a specific attack:

- first, the *Oracle attack*, also known as the sensitivity attack, defines a security scenario where the adversary has access to the watermark detector. In such a scenario, he's able to modify the watermarked content and to submit the modifications to the detector at will. This scenario can practically correspond to a playing device calling an access control system relying on the watermark detection. By analysing the answers of the detector (the Oracle) the adversary can find a strategy to both estimate the detection region and to move attacked contents outside the detection region while minimising the distortion.

Other scenarios assume that the adversary observes a set of contents watermarked using the same secret key.

- The *Known Message Attack* (a.k.a. KMA) assumes that the adversary observes a set of pairs associating embedded messages and watermarked contents. For example, it means that the adversary knows that his name or his identification number is embedded inside contents that he has bought on a pay per view platform.
- The *Known Original Attack* (a.k.a. KOA) assumes that the adversary observes a set of pairs associating original contents and watermarked contents. This class of attack may be for example performed if the content delivery chain allows the

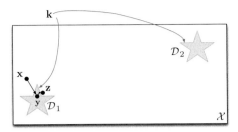

Fig. 1.2 Embedding and security attacks in watermarking: during embedding, the host vector $\mathbf{x}$ is moved into the secret region $\mathcal{D}_1$, generated using the secret key $\mathbf{k}$, to create the watermarked vector $\mathbf{y}$. If an adversary can estimate $\mathcal{D}_1$ or the secret key $\mathbf{k}$, he can then perform a "worst case attack", which means that the attacked content $\mathbf{z}$ is not detected as watermarked, and he can minimise the distortion between $\mathbf{y}$ and $\mathbf{z}$

broadcast- for example on TV channels or for teasers- of non watermarked contents.

- The *Constant Message Attack* (a.k.a. CMA) assumes that the contents are watermarked using the same message. This attack can be seen as a refinement of the KOA.
- The *Watermarked contents Only Attack* (WOA) assumes that the adversary observes only a collection of watermarked contents. Among all these scenario, this is the one where the adversary's knowledge is minimal.

These different scenarios allow us to specify the material available to the adversary to design his attack. Depending on this material, the adversary's goal can be to perform a worst case attack, consisting in preventing the detection or the decoding of the watermark while minimising the attacking distortion. This attack can be deterministic if the adversary is able to estimate all the parameters related to the secret key, or it can be partly random if only one subset of the parameters are estimated.

For illustration purposes, the international challenge BOWS-2[13] consisted in benchmarking the security of a watermarking scheme w.r.t. two security scenarios. The main goal of BOWS-2 was to prevent the watermark detection on 3 images provided by the server, and to minimise the attack distortion. The first security scenario was dedicated to Oracle attacks, participants having an unlimited access to the server, and the winners have called the detector several millions of times in order to reach a *PSNR* of 50 dB. A second security scenario was related to *information leakage* attacks, the number of submission was limited to 3 per day to prevent Oracle attacks, and participants were given 10,000 images watermarked with a same key. These participants were consequently able to perform a Constant Message Attack and the winner obtained a *PSNR* of 45 dB.

---

[13]http://bows2.ec-lille.fr/.

1.3.4 Security or Robustness?

The definition of security proposed in this document is rather simple: **we will talk about watermarking security whenever an adversary will be part of the game**. Consequently the action consisting in damaging the main functionality of the system will be called an **attack**. On the other hand, the word robustness will be used to define the usual processes that the content will encounter during its digital life. An additive noise, a convolution with a filter, or the translation of a watermarked signal are as many processes that will be used to benchmark the robustness of a system and will be denoted usual processes.

The problem of watermarking security can be then defined as the occurrence of an adversary aiming to break the system. Contrary to other definitions of watermarking security [7, 10] defining the problem of security as the estimation of the secret, this definition attends to be more general. The estimation of the secret is only a possible consequence of the occurrence of the adversary, but not the only one. As an example, the "worst case attack" dedicated to QIM [11] or the Oracle attacks [8, 12] are security attacks which don't directly involve the estimation of the secret key (Fig. 1.3).

To conclude, it is important to note that if the question "*what is the watermarking security?*" can seem easy, its answer doesn't bring any practical clue on how to conduct a security analysis. The more interesting question "*is the watermarking system X secure?*" is much more complex. As already mentioned in this introduction, the practical security analysis of a watermarking system must take into account different constraints such as the nature of the watermarked content, the watermarking scheme, the adversary's materials, his goals, or his power (see Fig. 1.4).

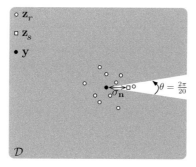

Fig. 1.3 Differences between an attack (linked with the definition of security) and an usual process (linked with the definition of robustness). The content is detected as watermarked if it belongs to the grey area $\mathcal{D}$. If the usual process is an additive Gaussian noise $\mathbf{n}$, $\mathbf{n} \sim \mathcal{N}(0, \sigma_{\mathbf{n}}^2 \mathbf{I})$ such that $\mathbf{z}_r = \mathbf{y} + \mathbf{n}$ (the AWGN channel), the probability of missed detection is $p_{\mathrm{md}} = \theta/2\pi = 5 \times 10^{-2}$. On the other hand, if the adversary knows $\mathcal{D}$, his attack leads to $p_{\mathrm{md}} = 1$ for a same distortion $||\mathbf{n}||$ than the usual process. After such an attack the content $\mathbf{z}_s$ will be outside $\mathcal{D}$

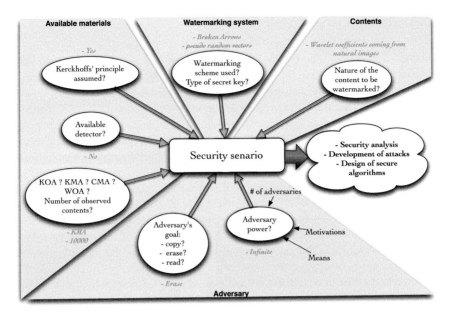

Fig. 1.4 The methodology used in watermarking security. In order to perform a security analysis, one needs to define accurately the different ingredients of the security scenario. The cursive text illustrates ingredients which have been used for the third episode of the BOWS-2 contest

1.3.5   The Stakes Behind Watermarking Security

By taking into account the framework defined in this section, we can now define what are the different problems that arise when we link watermarking with security. Among those, three will be of interest in the sequel of this book.

1. **How to define the security of a watermarking scheme?** We have seen in this section that we can define different watermarking scenarios, but are there also different classes of attacks? Another related question is how to define mathematically the security of a watermarking scheme?
2. **Is secure embedding possible? If so, what is the robustness of such as scheme?** It is important here to define schemes that belong to a given security class and to study the interplay between the security constraint and the two other constraints of watermarking which are robustness and imperceptibility.
3. **What are the watermarking schemes which are not secure? What is their security level? What kind of practical attacks can be set up?** The goal here will be to study the security of different popular watermarking schemes and to list different tools that are needed to perform a security attack.

References

1. Verance (2010) Cinavia Technologie. http://www.cinavia.com
2. Prigent N, Karroumi M, Chauvin M, Eluard M, Maetz Y, France TRD (2007) AACS, nouveau standard de protection des contenus pré-enregistrés haute-définition
3. Jin H, Lotspiech J (2008) Renewable traitor tracing: a trace-revoke-trace system for anonymous attack. Comput Secur-ESORICS 2007:563–577
4. Nakashima Y, Tachibana R, Babaguchi N (2009) Watermarked movie soundtrack finds the position of the camcorder in a theater. IEEE Trans Multimed 11(3):443–454.ISSN 1520-9210
5. Benichou D (2009) The seven cybercriminal families (invited speaker). In: IEEE international workshop on information forensics and security
6. Grimes R (2011) Your guide to the seven types of malicious hackers. http://www.infoworld.com/d/security-central/your-guide-the-seven-types-malicious-hackers-636?source=IFWNLE_nlt_sec_2011-02-08. Feb 2011
7. Cayre F, Fontaine C, Furon T (2005) Watermarking security: theory and practice. IEEE Trans Signal Process. Special issue Supplement on Secure Media II 53:3976–3987 (2005)
8. Linnartz J-P, van Dijk M (1998) Analysis of the sensitivity attack against electronic watermarks in images. In: International information hiding workshop
9. Pérez-Freire L, Pérez-González F, Furon T (2006) On achievable security levels for lattice data hiding in the known message attack scenario. In: 8th ACM multimedia and security workshop. Geneva, Switzerland, pp 68–79, Sep 2006 (accepted)
10. Pérez-Freire L, Pérez-Gonzalez F (2009) Spread spectrum watermarking security. IEEE Trans Inf Forensics Secur 4(1):2–24
11. Vila-Forcén JE, Voloshynovskiy S, Koval O, Pérez-González F, Pun T (2005) Worst case additive attack against quantization-based data-hiding methods. In: Edward J Delp III, Wong PW (eds) Security, steganography, and watermarking of multimedia contents VII. San Jose, California, USA: SPIE, Jan 2005
12. Comesaña P, Freire LP, Pérez-González F (2006) Blind Newton sensitivity attack. IEE Proc Inf Secur 153(3):115–125

Chapter 2
A Quick Tour of Watermarking Techniques

In order to understand and analyse the main components of watermarking security presented in the next chapters, we introduce in this chapter the different elements needed to embed a watermark or a message inside a host content. We first present a functional view of a watermarking scheme (the embedding function, the decoding/detection function) and then its geometrical interpretation. Then we present the most popular class of watermarking schemes: spread-spectrum watermarking and watermarking techniques based on the idea of dirty paper codes.

2.1 Basic Principles

2.1.1 Watermarking Constraints

As presented in Fig. (1.3), the three main constraints of watermarking are security, imperceptibility and capacity. Imperceptibility measures the distortion between host contents and watermarked contents, it can be measured using classical l-norm distances or perceptive measures derived from the human system. Capacity measures the amount of information that can be transmitted by the watermarking system through the channel, it can be computed using information theoretical measures such as the mutual information. Robustness can be seen as a practical translation of capacity and can be measured by the symbol error rate through the watermarking channel. In order to have a practical watermarking system whose performances are close to capacity, a system has to use an appropriate channel coding system. The last constraint, security is more complex and defined deeply within this book, particularly in Chap. 3. Figure 2.1 illustrates two triangles of constraints, a theoretical one in black takes into account capacity, and a practical one in red takes into account robustness.

© The Author(s) 2016
P. Bas et al., *Watermarking Security*, SpringerBriefs in Signal Processing,
DOI 10.1007/978-981-10-0506-0_2

Fig. 2.1 The different
constraints of Watermarking

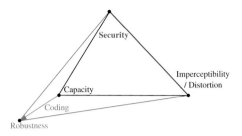

2.1.2   Zero-Bit and Multi-bit Watermarking

Watermarking schemes are divided into two classes called zero-bit and multi-bit
watermarking.

For zero-bit watermarking, a watermark is embedded in a host content and the goal
is to detect if a content is watermarked or not. Watermark detection can therefore be
seen as an hypothesis test with two hypotheses: H_0 : "content no watermarked" and
H_1 : "content watermarked". The performance of a zero-bit watermarking scheme is
evaluated regarding the false alarm probability or false detection probability p_{fd} (i.e.
the probability to detect a watermark in a non-watermarked content) and the miss
detection probability (i.e. the probability to not detect a watermark in a watermarked
content).

For multi-bit watermarking, a message m is embedded in a host content and the
goal is to decode the embedded message. The performance of a multi-bit watermarking scheme is evaluated with respect to the message error rate, i.e. probability
to wrongly decode a message, or the bit error rate, i.e. the probability to decode
wrongly embedded bits.

We now present two equivalent formalisations of watermarking, the first one is the
processing view, where both the embedding and the decoding schemes are seen as
processes or functions. The second view is more geometrical and allows us to picture
effects of embedding, decoding and also the role of the secret key in Watermarking.

2.1.3   A Processing View

The main chain of processes is the following: from a content c_x is extracted an host
signal $x \in X$ which is transformed into a watermarked content $y \in X$ using a secret
key $k \in K$.

The extraction process $\psi(.)$ is public (it does not depend of a secret key) and
invertible, $x = \psi(c_x)$. It can be for example the Least Significant Bits of an image,
its Discrete Cosine Transform coefficients or Wavelets subbands. An inverse function
$\psi^{-1}(.)$ is used to generate the watermarked content $c_y = {}^{-1}(y, \bar{x})$, where $\bar{x}$ represents
the components of c_x which are not present in x. Because watermarking processes

can be used on any signal $\mathbf{x}$, input signal $\mathbf{c_x}$ and output signal $\mathbf{c_y}$ will be omitted in the rest of this chapter.

A message m is embedded, if m is a power of 2, it can be decomposed into a set of N_m bits $\{b_1, \ldots, b_{N_m}\}$.

The embedding is performed using an embedding function $e(.)$ to generate the watermarked content:

$$\mathbf{y} = e(\mathbf{x}, \mathbf{k}, m), \tag{2.1.1}$$

where $\mathbf{k}$ denotes the secret key. It enables for example to select a subset of coefficients of $\mathbf{c_x}$, or to generate a set of pseudo random vectors seeded with $\mathbf{k}$ used to build a secret subspace or to scramble the content space $\mathcal{X}$. This embedding can be seen as an additive process involving the watermark vector $\mathbf{w} = \mathbf{y} - \mathbf{x}$.

The decoded message is estimated using a decoding function

$$\hat{m} = d(\mathbf{z}, \mathbf{k}), \tag{2.1.2}$$

from a potentially corrupted vector

$$\mathbf{z} = \mathbf{y} + \mathbf{n}, \tag{2.1.3}$$

where $\mathbf{n}$ is an additive noise. We can unify multi-bit and zero-bit schemes by embedding a constant message $m = \mathrm{cst} = 1$ for zero-bit watermarking. The decoding function consequently becomes a detection function: $\hat{m} = d(\mathbf{z}, \mathbf{k}) = 1$ if the watermark is detected and $\hat{m} = d(\mathbf{z}, \mathbf{k}) = 0$ if not.

In this chain of processes, we can highlight the watermarking channel which converts the message m into a watermarked content $\mathbf{y}$ and decodes the message $\hat{m}$ from the corrupted watermarked content $\mathbf{z}$. This channel is private and accessible using the secret key $\mathbf{k}$ or as we will see in Sect. 3.3 perturbed versions of $\mathbf{k}$. The embedding and decoding functions are considered as public (this is the Kerckhoffs' principle which states that the security relies only on the usage of a secret key) and one goal of the adversary is to have access to the watermarking channel. By doing so, the adversary is able to embed a new message, to alter the current one, to copy it into another content, or to decode it from a watermarked content. The access to the watermarking channel can be total (a read and write access) or only partial (the possibility to copy the message into another content, or to alter it).

2.1.4 The Geometrical View

A more geometrical view of the effects of embedding and decoding in the content space $\mathcal{X}$ is now presented. This view is motivated by the decoding function $d(.)$ which induces two entities:

Table 2.1 Equivalences between processing and the geometrical views

Processing view	Geometrical view
Embedding function $e(.)$	Choice of the watermark vector $\mathbf{w}$
Embedded message m	Subset of $\mathcal{D}$ having label m
Decoding function $d(.)$	Identification of the label $m(i)$ such that $\mathbf{z} \in \mathcal{D}_i$
Secret key $\mathbf{k}$	Set of decoding regions $\mathcal{D}$ and the labelling function $m(.)$

1. a partition of $\mathcal{X}$ into a set of N_d decoding regions $\mathcal{D} = \{\mathcal{D}_1, \ldots, \mathcal{D}_{N_d}\}$, N_d can be infinite,
2. a labelling function $m(.)$ which is a surjective function from $[1 .. N_d]$ to $[1 .. N_m]$ (hence $N_d \geq N_m$).

Each decoding region $\mathcal{D}_i$ is labeled by a message $m_i = m(i)$ and the function $d(.)$ assigns to $\mathbf{z}$ the label of the decoding region it belongs to:

$$\hat{m} = m(i) \text{ if } \mathbf{z} \in \mathcal{D}_i. \tag{2.1.4}$$

Note that in the case of zero-bit watermarking, there is only two possible labels $m(i) = 0$ or $m(i) = 1$. In this case, the content $\mathbf{z}$ is detected as watermarked only if $\hat{m} = 1$ and $\hat{m} = 0$ means that the content is detected as original.

The set $\mathcal{D}$ can be generated from the Voronoï cells of N_d points of $\mathcal{X}$ (see for example Sects. 2.4 and 2.4.3) or by the boundaries of the decoding regions (see for example Sects. 2.2 and 2.3). In order to minimise the embedding distortion, the embedding function $e(.)$ in going to "push" the host content $\mathbf{x}$ into the closest decoding region $\mathcal{D}_j$ labeled with the message m to embed (i.e. satisfying $m = m(j)$) by adding a properly chosen watermark vector $\mathbf{w}$ to obtain the watermarked content $\mathbf{y} = e(\mathbf{x}, \mathbf{k}, m) = \mathbf{x} + \mathbf{w}$. The equivalence between the elements of the processing view and those of the geometrical view are summed up in Table 2.1 (Fig. 2.2).

2.1.5 Three Fundamental Constraints

A watermarking scheme has also to take into account three fundamental constraints presented below.

The **Robustness**, which can be defined as the probability to correctly decode the embedded message from $\mathbf{z}$. Usually, the probability to wrongly decode the message is used instead and is computed using the message error rate $p_{me} = E_M[\mathbb{P}(\hat{M} \neq M)]$ or the bit error rate $p_{be} = E_B[\mathbb{P}(\hat{B} \neq B)]$ when the message is composed of a set of embedded bits. The goal of the watermarking designer will be to maximise the robustness, i.e. to minimise the probability of error. In order to do so, the embedding function will try to push the watermarked content inside the appropriate decoding region. The farther $\mathbf{y}$ is from the boundaries of $\mathcal{D}_i$ the more important is the robustness.

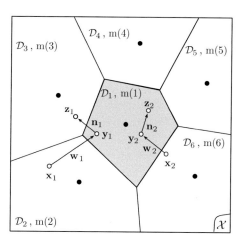

Fig. 2.2 A geometrical view of watermarking: in this example host vectors $\mathbf{x}_1$ and $\mathbf{x}_2$ are used to embed the same arbitrary message $m = \mathrm{m}(1)$ and consequently have to be moved in region $\mathcal{D}_1$. Symbol • represents centres of the Voronoï cells defining the decoding regions, and ○ represents contents in the content space $\mathcal{X}$. An embedding function generates two watermark vectors $\mathbf{w}_1$ and $\mathbf{w}_2$ which are added to create the watermarked contents $\mathbf{y}_1$ and $\mathbf{y}_2$ respectively. In this example, the first embedding is less robust than the second one because $\mathbf{y}_1$ is closer to the boundary of $\mathcal{D}_1$ than $\mathbf{y}_2$. The message error rate will be consequently more important for the first content than for the second one (for a same Watermark to Noise Ratio under the AWGN channel). The noise $\mathbf{n}_1$ generates a decoding error but not the noise $\mathbf{n}_2$

The probability of decoding error is usually computed for a given Watermark to Noise Ratio, defined as the WNR $= \log_{10}\left(\mathrm{E}[W^2]/\mathrm{E}[N^2]\right)$ dB. Note also that another important and popular constraint, the **Capacity**, can be seen as the dual constraint of Robustness: for a given channel, the watermarking designer wants either to reach the capacity in order to transmit as much information as possible; or for a given transmission rate, he wants to maximise robustness, i.e. to minimise the message error rate. In each case however the watermarking designer will have to find the best code given his objective, the code maximising robustness for a given channel will be the same than the code maximising the transmission rate, i.e. reaching the capacity.

The **Distortion**, classically defined as the power of the watermark signal $\mathrm{E}[W^2]$ or the Document to Watermark Ratio DWR $= \log_{10}\left(\mathrm{E}[X^2]/\mathrm{E}[W^2]\right)$ dB and is equal to DWR $= 10 \cdot \log_{10}\left(\sigma_x^2/\sigma_w^2\right)$ dB if both signals are centred. Note that the distortion measure can be more elaborated in order to take into account psychovisual criterions. The goal of the watermarking designer will be to minimise the distortion or to set it below a threshold. This means that the distance between $\mathbf{x}$ and $\mathbf{y}$, or its expectation, has to be bounded or minimised. The same rule applies for the distance between between $\mathbf{x}$ and the different decoding regions.

The **Security**, defined by Kalker as the inability for the adversary to have access to the watermarking channel [1], which can be translated as the inability to estimate the secret key $\mathbf{k}$ (or an approximation of it) granting the reading or writing access of the message. Under the geometrical viewpoint, $\mathbf{k}$ allows us to generate the set $\mathcal{D}$ and the labelling function m(.). We can see that if the adversary knowns these two entities, he is able to have access to the watermarking channel, e.g. he can embed or decode messages. This last constraint will motivate the next chapters of this document and will be formalised in the next chapter.

Note that the key $\mathbf{k}$ which will be used in the following of the document is generated from a user key $\mathbf{k}_u$ which is defined by the user who wants to embed of decode the message. The key which is used to have access to the watermarking channel in the content space $\mathcal{X}$, it not $\mathbf{k}_u$ but $\mathbf{k}$ which represents a set of signals and parameters that are generated from $\mathbf{k}_u$. From a security point of view, finding the user key $\mathbf{k}_u$ is equivalent to finding $\mathbf{k}$ or the set $\{\mathcal{D}, \mathrm{m}(.)\}$: each entity allows us to have a complete access to the watermarking channel. We consequently have the following relation $\mathbf{k}_u \rightarrow \mathbf{k} \rightarrow \{\mathcal{D}, \mathrm{m}(.)\}$. In the sequel, we will omit to define the user key since it is related to the implementation of the watermarking system, and it classically corresponds to the seed of a pseudo-random generator.

It is also important to note that these three constraints are strongly dependant from each other. For example, the robustness grows when the decoding regions become large and the contents goes deep inside one of them; however such a strategy implies that the distortion grows as well. A balance between distortion and robustness has consequently to be found. We will see in the next chapter that the same applies regarding the security.

We now present a quick tour of the most popular watermarking schemes and we detail the principles of embedding, decoding or detection functions.

2.2 Spread Spectrum and Improved Spread Spectrum

Spread Spectrum Watermarking

Also abbreviated SS, it uses the same principles than Spread Spectrum communications to transmit a message in a noisy environment [2]. The bandwidth of the message is spread on a larger bandwidth thanks to a modulation with one or several carriers represented by pseudo-random vectors.

For this basic implementation, SS watermarking allows us to embed $N_b = \log_2(N_m)$ bits $(b_1, \ldots, b_{N_b})$. Under given security scenarios, security is granted thanks to the use of a set of N_m pseudo-random vectors $\mathbf{k} = \{\mathbf{k}_1, \ldots, \mathbf{k}_{N_m}\}$ seeded by the user key (see Sect. 2.1.5). Without loss of generality, we can normalise and orthogonalize each secret vector ($\forall(i \neq j), ||\mathbf{k}_i|| = 1$ and $\mathbf{k}_i'\mathbf{k}_j = 0$), and the embedding formula is:

$$\mathbf{y} = e(\mathbf{x}, m, \mathbf{k}) = \mathbf{x} + \alpha \sum_{i=1}^{N_b} (-1)^{b_i} \mathbf{k}_i. \tag{2.2.1}$$

The watermark vector is $\mathbf{w} = \alpha \sum_{i=1}^{N_b}(-1)^{b_i}\mathbf{k}_i$, and $||\mathbf{w}||^2 = N_b\alpha^2 = \text{cst.}$ Note that α is a scalar used to choose the embedding distortion DWR $= 10 \cdot \log_{10}\left(\frac{N_v\sigma_x^2}{N_b\alpha^2}\right)$ dB which means that

$$\alpha = \sqrt{N_v/N_b}\sigma_x 10^{-\text{DWR}/20}. \tag{2.2.2}$$

The decoding is performed on the potentially corrupted vector $\mathbf{z}$ by computing $\mathbf{z}^T\mathbf{k}_i$, i.e. the projections of the watermarked vector on each pseudo-random vector:

$$\hat{b}_i = 2 \cdot \text{sign}(\mathbf{z}^T\mathbf{k}_i) - 1. \tag{2.2.3}$$

The boundaries of the decoding regions are consequently defined by the N_b hyper plans orthogonal to each $\mathbf{k}_i$ which yields to 2^{N_b} different decoding regions, (one for each message). Figure 2.3 represents the decoding regions and the locations of watermarked contents projected on the plan $(O, \mathbf{k}_1, \mathbf{k}_2)$ for $N_b = 2$ and $(O, \mathbf{k}_1, \mathbf{u}_\perp)$ for $N_b = 1$ where $\mathbf{u}_\perp$ is a random unitary vector orthogonal to $\mathbf{k}_1$. Under an AWGN channel defined by a variance σ_n^2 and WNR $= 10\log_{10}\left(\alpha^2/\sigma_n^2\right)$, the BER is computed using the projection $\mathbf{Z}^T\mathbf{K}_i$ which is a Gaussian random variable of law $\mathcal{N}(\alpha, \sigma_x^2 + \sigma_n^2)$, and the BER is given by:

$$p_{be} = \frac{1}{2}\left[1 + \text{erf}\left(\frac{-\alpha}{\sqrt{2\left(\sigma_x^2 + \sigma_n^2\right)}}\right)\right] = \frac{1}{2}\left[1 + \text{erf}\left(-\sqrt{\frac{N_v 10^{-\text{DWR}/10}}{2N_b\left(1 + 10^{-(\text{WNR}+\text{DWR})/10}\right)}}\right)\right]. \tag{2.2.4}$$

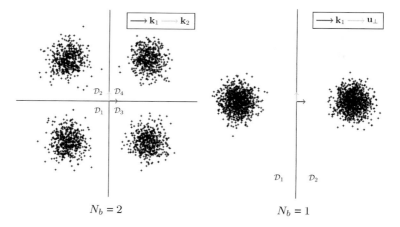

Fig. 2.3 Geometrical view of SS watermarking—Watermarked contents projected on $\mathbf{k}_1$ and $\mathbf{k}_2$ ($N_b = 2$) or $\mathbf{u}_\perp$ ($N_b = 1$) and the associated decoding regions. $N_v = 128$, Gaussian host, DWR $= 5$ dB

2.2.1   Improved Spread Spectrum

Improved Spread Spectrum (ISS) watermarking [3] is a variant of Spread Spectrum and is motivated by the fact that for SS the embedding (2.2.1) uses only one parameter α. All host contents are consequently moved by a watermark vector $\mathbf{w}$ of constant norm, only the direction depends on the message to embed. In order to take into account the robustness constraint, the authors of [3] have proposed to use a second parameter γ which is tuned in order to minimise the bit error rate for a given WNR under an AWGN channel. The term $\gamma \in [0, 1]$ is chosen to perform **host rejection**, which consists in reducing the interference (or intercorrelation) $\mathbf{x}^T \mathbf{k}_i$ between the host $\mathbf{x}$ and each pseudo-random vector $\mathbf{k}_i$ by adding the term $-\gamma \left(\mathbf{x}^T \mathbf{k}_i \right) \mathbf{k}_i$, and it is equivalent to reducing the variance of the watermarked contents along each $\mathbf{k}_i$. Once the variance is reduced, the content is moved using a constant vector, proportional to $\mathbf{k}_i$ using a parameter α similar to the one used in Spread Spectrum:

$$\mathbf{y} = e(\mathbf{x}, m, \mathbf{k}) = \mathbf{x} + \sum_{i=1}^{N_b} \left(\alpha(-1)^{m_i} - \gamma \mathbf{x}^T \mathbf{k}_i \right) \mathbf{k}_i. \tag{2.2.5}$$

The watermark vector is consequently $\mathbf{w} = \sum_{i=1}^{N_b} \left(\alpha(-1)^{m_i} - \gamma \mathbf{x}^T \mathbf{k}_i \right) \mathbf{k}_i$ and $||\mathbf{w}||^2 = N_b \left(\alpha^2 + \gamma^2 \sigma_x^2 \right)$ which gives

$$\text{DWR} = 10 \cdot \log_{10} \left(\frac{N_v \sigma_x^2}{N_b \left(\alpha^2 + \gamma^2 \sigma_x^2 \right)} \right) \text{ dB} \tag{2.2.6}$$

For a given γ this means that we have:

$$\alpha = \sqrt{ \left(\frac{N_v \sigma_x^2}{N_b} \right) 10^{-\text{DWR}/10} - \sigma_x^2 \gamma^2 } \text{ if } \left(\frac{N_v \sigma_x^2}{N_b} \right) 10^{-\text{DWR}/10} \geq \sigma_x^2 \gamma^2. \tag{2.2.7}$$

Note that for $\gamma = 0$, SS and ISS are identical. Figure 2.4 depicts the locations of watermarked contents using ISS embedding for $\gamma = 0.9$. We can notice that the host rejection strategy allows us to reduce the variance of the watermarked contents inside each decoding region, making them statistically less prone to leave the decoding region for a noise of relative small power than contents watermarked using SS.

The decoding of the message is performed exactly as for SS, i.e. using Eq. (2.2.3). The authors propose to compute γ in order to minimise the BER for an AWGN channel of given WNR. Under the AWGN, the projection $\mathbf{Z}^T \mathbf{K}_i$ is a Gaussian random variable of law $\mathcal{N}(\alpha, (1 - \gamma)^2 \sigma_x^2 + \sigma_n^2)$ and the BER is given by:

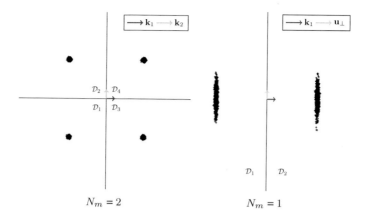

Fig. 2.4 Geometrical view of ISS watermarking—Watermarked contents projected on $\mathbf{k}_1$ and $\mathbf{k}_2$ ($N_b = 2$) or $\mathbf{u}_\perp$ ($N_b = 1$) and the associated decoding regions. $N_v = 128$, Gaussian host, DWR $= 5\,$dB, $\gamma = 0.9$

$$p_{be} = \frac{1}{2}\left[1 + \mathrm{erf}\left(-\sqrt{\frac{\left(\frac{N_v \sigma_x^2}{N_b}\right)10^{-\frac{\mathrm{DWR}}{10}} - \sigma_x^2 \gamma^2}{2\left((1-\gamma)^2 \sigma_x^2 + \sigma_n^2\right)}}\right)\right],$$

$$= \frac{1}{2}\left[1 + \mathrm{erf}\left(-\sqrt{\frac{1}{2}\frac{\left(\frac{N_v}{N_b}\right)10^{-\frac{\mathrm{DWR}}{10}} - \gamma^2}{(1-\gamma)^2 + 10^{-\frac{\mathrm{WNR}+\mathrm{DWR}}{10}}}}\right)\right]. \tag{2.2.8}$$

The value $\gamma^{(R)}$ which minimises BER is given by solving $\partial p_{be}/\partial \gamma = 0$ and gives

$$\gamma^{(R)} = \frac{1}{2}\left((1 + b + a) - \sqrt{(1 + b + a)^2 - 4a}\right), \tag{2.2.9}$$

with $a = (N_v/N_b) \cdot 10^{-\mathrm{DWR}/10}$ and $b = \sigma_n^2/\sigma_x^2 = 10^{-(\mathrm{WNR}+\mathrm{DWR})/10}$.

The comparison between the two BERs is illustrated on Fig. 2.5, we can notice that when the noise power increases, the performance of ISS embedding tends to be the

Fig. 2.5 Comparison between the BER for ISS and SS w.r.t. the WNR ($N_v = 64$, $N_b = 8$, DWR $= 5\,$dB)

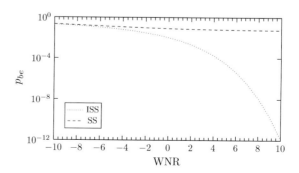

same than the performance of SS embedding. This is due to the fact that $\gamma^{(R)} \to 0$ when WNR $\to -\infty$. On the other hand, the improvement of ISS regarding SS is really important for low WNRs, i.e. when the host rejection is important and $\gamma^{(R)} \to 1$.

2.3 Correlation Based Zero-Bit Watermarking

The principle of spread spectrum watermarking can also be used for zero-bit watermarking. As for SS decoding, the detection is also performed using the correlation $\mathbf{z}^T \mathbf{k}$ between the secret key $\mathbf{k}$ and the watermarked (and possibly corrupted) content $\mathbf{z}$. However, since the watermark has to be detected also on possibly scaled versions of the watermarked contents $\mathbf{z} = \alpha \mathbf{y}$, the detection is performed by computing the normalised correlation instead:

$$d(\mathbf{z}, \mathbf{k}) = 1 \quad \text{if} \quad \frac{|\mathbf{z}^T \mathbf{k}|}{||\mathbf{z}^T|| \cdot ||\mathbf{k}||} > T,$$
$$d(\mathbf{z}, \mathbf{k}) = 0 \text{ else.} \tag{2.3.1}$$

Note that $|\mathbf{z}^T \mathbf{k}| / \left(||\mathbf{z}^T|| \cdot ||\mathbf{k}|| \right) = |\cos \theta|$ where θ is defined as the angle between the two vectors $\mathbf{k}$ and $\mathbf{z}$. This means that the boundary of the detection region is defined by a double hyper-cone (called here $\mathcal{C}$) of axis $\mathbf{k}$ and angle $\theta = \arccos(T)$. The threshold T (or θ) is computed in order to satisfy a given false alarm probability $\mathbb{P}[d(X, K) = 1] = p_{fa}$. Since K is uniformly distributed on an hyper sphere of dimension N_v, this probability can be computed as the ratio between twice the surface of the hyper-cap of solid angle θ over the surface of the hyper-sphere [4]:

$$p_{fa} = 1 - I_{\cos^2 \theta} \left(1/2, (N_v - 1)/2 \right), \tag{2.3.2}$$

where $I(.)$ denotes the regularised incomplete beta function.

Without loss of generality we set $||\mathbf{k}|| = 1$. Once the decoding region is defined, the embedding has to be designed in order to move the host content $\mathbf{x}$ into $\mathcal{C}$. In order to find the adequate subspace to choose $\mathbf{w}$, [5] proposed to build a plan $(O, \mathbf{k}, \mathbf{e}_2)$ defined by the secret key $\mathbf{k}$ (the cones axis) and the unitary vector $\mathbf{e}_2$ orthogonal to $\mathbf{k}$, such that the host vector belongs to the plan $(O, \mathbf{k}, \mathbf{e}_2)$:

$$\mathbf{e}_2 = \frac{\mathbf{x} - \left(\mathbf{x}^T \mathbf{k} \right) \mathbf{k}}{||\mathbf{x} - \left(\mathbf{x}^T \mathbf{k} \right) \mathbf{k}||}. \tag{2.3.3}$$

Since the closest point to $\mathbf{x}$ included in $\mathcal{C}$ belongs to $(O, \mathbf{k}, \mathbf{e}_2)$, this plan spans the shortest direction to enter inside the cone.

Different strategies are possible in order to choose the watermark vector $\mathbf{w}$. For all these strategies, we assume that the norm of the watermark vector is constant and equal to $D = \sqrt{N_v \sigma_x^2 \cdot 10^{-\text{DWR}/10}}$.

In [5], the authors propose to assume that the noise $\mathbf{n}$ and the watermarked content $\mathbf{y}$ are orthogonal ($\mathbf{y}^T\mathbf{n} = 0$), this assumption is more and more realistic when N_v grows and the content suffers an AWGN channel. The goal here is to find the point inside $\mathcal{C}$ such that the distance R between $\mathbf{y}$ and the cone boundary considering a direction orthogonal $\mathbf{y}$ is maximal. Thanks to the Pythagorus theorem,

$$R = \sqrt{(\mathbf{y}[1]\tan\theta)^2 - \mathbf{y}[2]^2}, \tag{2.3.4}$$

and we can use a basic optimisation technique to find $\mathbf{y}_{\mathrm{ON}}$ such that R is maximal (ON stands for Orthogonal Noise). See Fig. 2.6 for a geometrical illustration.

In [6] (see also Sect. 4.3 of this book), the authors propose to maximise the robustness considering the worst case scenario, i.e. the fact that the adversary knows the cone axis. In this setup, the robustness can be defined as the distance between $\mathbf{y}$ and the nearest point of the cone boundary. The solution to maximise this distance is to move $\mathbf{x}$ in a direction Orthogonal to the Boundary (hence the denomination OB). In certain cases the watermarked content can reach the cone axis, and because of the axial symmetry of the cone, the best strategy then is to go in the same direction than $\mathbf{k}$ once the axis is reached. The embedding formulae is literal and can be written as:

$$\text{if } D < \frac{\mathbf{x}[2]}{\cos\theta} : \begin{cases} \mathbf{y}_{\mathrm{OB}}[1] = \mathbf{x}[1] + D\cdot\sin\theta \\ \mathbf{y}_{\mathrm{OB}}[2] = \mathbf{x}[2] - D\cdot\cos\theta \end{cases} \tag{2.3.5}$$

and

$$\text{if } D \geq \frac{\mathbf{x}[2]}{\cos\theta} : \begin{cases} \mathbf{y}_{\mathrm{OB}}[1] = \mathbf{x}[1] + \sqrt{D^2 - \mathbf{x}[2]^2} \\ \mathbf{y}_{\mathrm{OB}}[2] = 0 \end{cases}. \tag{2.3.6}$$

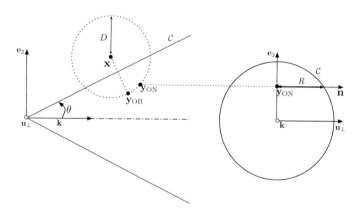

Fig. 2.6 *Left* geometrical presentations of the 3 different embedding strategies, Orthogonal Noise ($\mathbf{y}_{\mathrm{ON}}$), Orthogonal to the Boundary ($\mathbf{y}_{\mathrm{OB}}$) and by Maximizing the Mutual Information ($\mathbf{y}_{\mathrm{MMI}}$). *Right* computation of the robustness R when assuming that the noise is orthogonal to the watermarked content

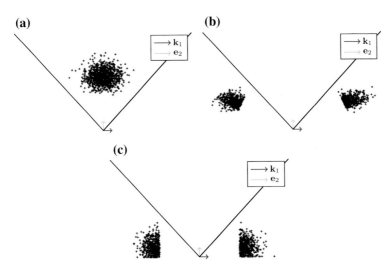

Fig. 2.7 Projection of the original contents, **a** and watermarked contents **b** orthogonal noise, **c** orthogonal to the boundary and **d** Max Mutual Information in the $(O, \mathbf{k}, \mathbf{e}_2)$. $N_v = 32$, DWR = 0 dB, $\theta = \pi/4$ ($p_{fa} = 4 \cdot 10^{-6}$)

The geometric illustration of the embedding "Orthogonal to the Boundary" is depicted on Fig. 2.6.

Figure 2.6 represents the geometrical interpretations of the two different embeddings in the plan $(O, \mathbf{k}, \mathbf{e}_2)$ (left) and in the plan plan $(O, \mathbf{u}_\perp, \mathbf{e}_2)$ for the first strategy (right).

Figure 2.7 show the projection of original and watermarked contents for the 3 different embedding, we can notice that only the second strategy brings few contents to the cone axis.

2.4   Watermarking Based on Dirty Paper Codes

If host rejection is one way to increase the robustness of the watermarking scheme by reducing the variance of the watermarked content inside the decoding regions, another strategy consists in generating several decoding regions for the same message. Such a strategy is inspired by the idea of "writing on dirty papers" proposed by Costa [7] who shows theoretically that for an AWGN channel, the capacity of the watermarking system solely depends on the variance of the noise and the embedding distortion, and is not dependant of the host signal (e.g. the "dirty paper").

The practical construction of such a "dirty paper" scheme relies on the generation of M bins, each bin containing codewords. For a host content $\mathbf{x}$ and a message to embed m, the embedding distortion is minimised by considering only the mth bin and going toward the nearest codeword of this bin. Such a strategy is also called *informed coding* [8] since the embedding is informed by taking into account the host state **x** and picking the right codeword that codes the message to embed *m*.

Under the geometric view, it means that if the host vector **x** is too far for one embedding region, the strategy proposed by Costa's paper is to select another embedding region labeled by the same message but closer to **x**, the number of embedding regions being limited by the robustness constraint.

Practically, the implementation of dirty paper watermarking implies several requirements:

- a function to generate the N_m bins in $\mathcal{X}$, which corresponds to the labelling function defined in Sect. 2.1.4,
- a function to generate the N_c codewords in each bin and the associated set of decoding regions $\mathcal{D}$ (with N_c possibly equal to $+\infty$),
- the use of a secret key **k** which governs the location of the decoding regions and prevents the access to the watermarking channel.
- If N_m and N_c are large, one has also to find a fast and efficient way to have access to the different bins, codewords and the set $\mathcal{D}$ into a potential high dimensional space $\mathcal{X}$. As we shall see in the next section, the use of lattices is a convenient way to generate bins and decoding regions, and quantisers associated to lattices can be used to perform embedding by efficiently picking the closest codewords (see next subsection). Another convenient way to have access to a high dimensional dictionary is to use trellis (see Sect. 2.4.3).

2.4.1 Distortion Compensated Quantisation Index Modulation (DC-QIM)

Watermarking using Distortion Compensated Quantisation Index Modulation (DC-QIM) was first proposed by Chen and Wornell [9] and has been extended for cubic lattices by Eggers et al. [10] and other classes of lattices by Moulin and Koetter [11].

One bin in Costa's framework is provided by a coarse lattice Λ_c representing a set of points in $\mathcal{X}$. The coarse lattice Λ_c can be defined as

$$\Lambda_c = \{\boldsymbol{\lambda} = G \cdot \mathbf{i} : \mathbf{i} \in \mathbb{Z}^{N_v}\}, \tag{2.4.1}$$

where G is a $N_v \times N_v$ generator matrix and defines a set of point in $\mathcal{X} = \mathbb{R}^{N_v}$. The N_m different bins are generated by using translations of Λ_c. The bin/lattice Λ_m coding the message *m* is given by $(\Lambda_m = \Lambda_c + \mathbf{d}_m + \mathbf{k})$ where $\mathbf{d}_m$ is called a coset leader and **k** is a secret key that allows us to obtain a secret lattice configuration (see Fig. 2.8 for an illustration of these entities) . The coset leaders $\{\mathbf{d}_m\}$ are constructed using a specific lattice construction [12–14]. The most popular are the self-similar constructions, which consists in applying a scaling factor (in this case $\Lambda_f = \alpha\Lambda_c$) and/or a rotation of the coarse lattice Λ_c. "Construction A" consists in generating the set $\{\mathbf{d}_m\}$ inside a unitary hypercube and then shifting the elementary construction over $\mathcal{X}$.

Fig. 2.8 Lattices/Bins for the different messages for $N_m = 3$ and $N_v = 2$ using the coset leaders $\mathbf{d}_{m=2}$ and $\mathbf{d}_{m=3}$ ($\mathbf{d}_{m=1} = \mathbf{0}$). In this example we set $\mathbf{k} = \mathbf{0}$. $\mathbf{g}_1$ and $\mathbf{g}_2$ are the basis vectors of the generator matrix used to generate Λ_c, and $\mathcal{D}_{m=1}$, $\mathcal{D}_{m=2}$, $\mathcal{D}_{m=3}$ represent three decoding regions

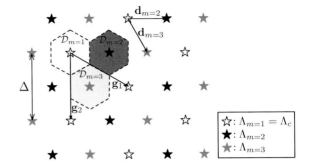

The set $\mathcal{D}$ of decoding regions associated to the watermarking scheme are the voronoï cells of the fine lattice Λ_f representing the union of the bins coding all the messages: $\Lambda_f = \cup_{m=1}^{N_m} \Lambda_m$:

$$\mathcal{D} = V(\Lambda_f), \qquad (2.4.2)$$

where the function $V(.)$ returns the Voronoï cell of each point of the lattice Λ_f.

One practical way to compute efficiently the nearest decoding region of one point $\mathbf{a}$ associated to the message m is to use a quantiser associated to the bin $\Lambda_m = \Lambda_c + \mathbf{d}_m + \mathbf{k}$ which is equivalent to quantise $\mathbf{a} - \mathbf{d}_m - \mathbf{k}$ according to Λ_c. The first implementation of Quantisation Index Modulation (QIM) simply proposes to embed a message m by quantising the host $\mathbf{x}$ using the appropriate bin Λ_m [9]:

$$e_{QIM}(\mathbf{x}, m, \mathbf{k}) = Q_{\Lambda_c}(\mathbf{x} - \mathbf{d}_m - \mathbf{k}) + \mathbf{d}_m + \mathbf{k}, \qquad (2.4.3)$$

the distortion being dependant of the quantisation step Δ, i.e. the distance between two closest elements of Λ_m. Note that the implementation of the quantisation can be straightforward or not, it depends of the lattice Λ_c.

The authors also propose to perform embedding by moving toward the closest point of Λ_m, i.e. moving inside the closest decoding region, using a distortion compensation parameter α. The method is called Distortion Compensated Quantisation Index Modulation (DC-QIM) and the embedding formula is given by:

$$e_{DC-QIM}(\mathbf{x}, m, \mathbf{k}) = \mathbf{x} + \alpha \left(Q_{\Lambda_c}(\mathbf{x} - \mathbf{d}_m - \mathbf{k}) - (\mathbf{x} - \mathbf{d}_m - \mathbf{k}) \right), \qquad (2.4.4)$$

which is identical to $e_{QIM}(.)$ for $\alpha = 1$. Note that depending on the lattice, there exists a minimum value α_{min} for which all the host contents are moved inside the appropriate decoding region. For embeddings using values below α_{min}, all the watermarked contents do not embed the correct message since the embedding distortion is not important enough (for $\alpha = 0$ the embedding in completely ineffective since $\mathbf{y} = \mathbf{x}$).

For both QIM and DC-QIM, the decoding function is the same, it consists in finding the bin $\Lambda_{m'}$ which induces the decoding region containing the corrupted vector $\mathbf{z}$ (see Fig. 2.8). Practically this is performed by computing the quantised value of $\mathbf{z}$ for each bin and by selecting the bin which provides the smallest quantisation error:

$$\hat{m} = d(\mathbf{z}, \mathbf{k}) = \arg\min_{m'} ||Q_{\Lambda_c}(\mathbf{z} - \mathbf{d}_{m'} - \mathbf{k}) - (\mathbf{z} - \mathbf{d}_{m'} - \mathbf{k})||. \qquad (2.4.5)$$

2.4.2 Scalar Costa Scheme (SCS)

Eggers et al. [10] have analysed the robustness performance of DC-QIM using a cubic lattice, i.e. using a uniform quantizer applied on each component. The quantisation step being Δ, we have for each component x of $\mathbf{x}$:

$$Q_{\Lambda_c}(x) = \text{sign}(x) \cdot \Delta \cdot \lfloor \frac{|x|}{\Delta} + \frac{1}{2} \rfloor, \qquad (2.4.6)$$

with $\lfloor . \rfloor$ the floor function. Each coset leader is equal to a translation of $d_m = \Delta \frac{m}{N_m}$ and the embedding becomes:

$$e(x, m, k) = x + \alpha \left(Q_{\Lambda_c}\left(x - \Delta\frac{m}{N_m} - k \right) - \left(x - \Delta\frac{m}{N_m} - k \right) \right), \qquad (2.4.7)$$

the scalar k being the secret key picked in the interval $[0, \Delta)$. In this setting the distribution of the host signal x is considered as piecewise uniform, additionally the embedding distortion is very small regarding the host signal, e.g. $\sigma_w^2 \ll \sigma_x^2$.

The distortion of the embedding is given by:

$$\sigma_w^2 = \frac{\alpha^2 \Delta^2}{12}. \qquad (2.4.8)$$

Figure 2.9 represents the distribution of watermarked contents for QIM and SCS with a given α_r in 2D ($N_v = 2$) for binary embedding on each component ($N_m/N_v = 2$) for the same embedding distortion DWR = 20 dB. We can notice the concentration of the watermarked contents on the quantisation cells for QIM and inside a square of length $(1 - \alpha)\Delta$ for SCS. Because of the structure of the lattice, the decoding regions are cubic.

In order to maximise the robustness, the author compute the achievable rate R of the watermarking channel which is given by the mutual information between the attacked signal and the embedded symbol:

$$R = I(Z, M) = -\int_\Delta p_Z(z) \log_2 p_Z(z) dz + \frac{1}{N_m} \sum_{m=1}^{N_m} \int_\Delta p_Z(z|m) \log_2 p_Z(z|d) dz,$$

$$\qquad (2.4.9)$$

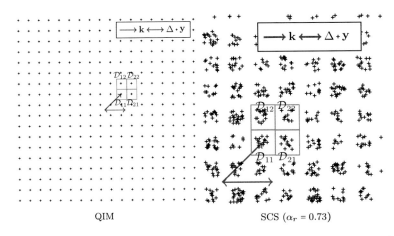

Fig. 2.9 Geometrical view of SCS watermarking for $N_v = 2$ and $N_m/N_v = 2$ (1 bit/sample) for QIM and SCS with 4 decoding regions . DWR = 20 dB

and the authors have derived an approximation of the embedding parameter maximising the achievable rate R for a given WNR. The approximation is given by [10]:

$$\alpha_r = \sqrt{\frac{1}{1 + 2.71 \cdot 10^{-\text{WNR}/10}}} . \qquad (2.4.10)$$

For this scheme, the minimum value of α which allows us to watermark contents without decoding error during the embedding is $\alpha_{min} = (N_m - 1)/N_m$. Figure 2.10 shows the robustness gap between QIM and SCS for an AWGN channel according to the WNR for a same embedding distortion (DWR = 20 dB), the Bit Error Rate p_{be} has been practically computed using 10^4 watermarked contents.

Fig. 2.10 Practical comparison between the BER for QIM and SCS w.r.t. the WNR ($N_v = 1, N_m = 2$, DWR = 20 dB) for an AWGN channel. The empirical probabilities are computed using 10^4 contents. ⤳ **implementation: run BERSCSPrac.R**

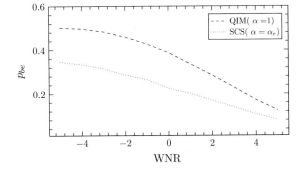

2.4.3 Trellis-Based Watermarking

The use of trellis for watermarking is a practical way to perform dirty paper coding [7]. Dirty paper coding implies the use of a codebook $\mathcal{C}$ of codewords with a mapping between codewords and messages. Dirty Paper Trellis (DPT) codes have two main assets: the generation of the codebook $\mathcal{C}$ is systematic and the search for the nearest codeword can be efficiently performed with a Viterbi decoder [15]. A DPT is a network defined by several parameters:

1. the number of states N_s,
2. the number of arcs per state N_a,
3. the N_v-dimensional pseudo-random patterns associated to each one of the $N_a \cdot N_s$ arcs, which can be assimilated to the carrier used in spread spectrum schemes,
4. the binary label associated to each one of the $N_a \cdot N_s$ arcs,
5. the number of steps in the trellis, corresponding in [8] to the number of bits N_b of the trellis,
6. the connectivity between the states, i.e. for each arc the labels of the two states which are connected.

Figure 2.11 depicts an example of a DPT. One can notice that the configuration of the trellis is simply repeated from one step to another. Moreover, the number of outgoing and incoming arcs per state is constant. These are common assumptions in trellis coding.

A DPT is thus associated with a codebook $\mathcal{C} = \{\mathbf{c_i}, i \in [1, \ldots, N_s \cdot N_a^{N_b}]\}$ of $N_s \cdot N_a^{N_b}$ codewords in a $N_v \cdot N_b$-dimensional space. Each codeword can be built in such a way that it corresponds to a path in the trellis and encodes N_b bits. This message can be retrieved by concatenating the binary labels of the arcs along the corresponding path.

DPT watermarking makes use of both *informed coding* and *informed embedding* [8]. Informed coding consists in selecting the codeword $\mathbf{g}$ in the codebook $\mathcal{C}$ that is the closest to the host vector $\mathbf{x}$ and that encodes the desired message. This is

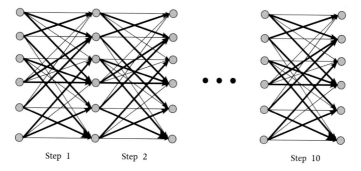

Step 1   Step 2   Step 10

Fig. 2.11 Example of the structure of a 10 steps trellis with 6 states and 4 arcs per states. *Bold* and normal arcs denote respectively 0 and 1 valued labels

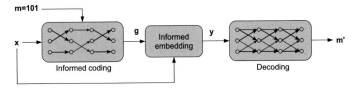

Fig. 2.12 Principles of informed embedding and informed coding using trellis codes

the same principle than for dirty paper coding. The selection is done here by running a Viterbi decoder with an expurgated trellis containing only arcs whose binary labels are in accordance with the message to be embedded. As a result, any path through the trellis encodes the desired message. The Viterbi decoder is then used to maximize/minimize a given function i.e. to find the *best* codeword in this subset according to some criterion. In their original article [8], the authors proposed to keep the codeword with the highest linear correlation with the host vector $\mathbf{x}$.

At this point, informed embedding is used to reduce the distance between the host vector $\mathbf{x}$ and the selected codeword $\mathbf{g}$. It basically computes a watermarked vector $\mathbf{y}$ that is as close as possible from $\mathbf{x}$ while being at the same time within the detection region of the desired codeword $\mathbf{g}$ with a guaranteed level of robustness to additive white Gaussian noise (AWGN). In practice, a sub-optimal iterative algorithm is used combined with a Monte-Carlo procedure to find this watermarked vector $\mathbf{y}$ [8].

On the receiver side, the embedded message is extracted by running a Vitterbi decoder with the whole DPT. The optimal path is thus identified and the corresponding message retrieved by concatenating the binary label of the arcs along this path. The whole procedure is illustrated in Fig. 2.12.

2.5 Conclusions of the Chapter

We can see from this chapter that, once the watermarking scheme is fully described, it is possible to compute the distortion and the robustness of the scheme, either using analytical formulas or by practically measuring the visual impact and the symbol error rate after a given process. In the next chapter we show how to measure the third constraint of the watermarking game, namely security, and how to distinguish different scheme w.r.t. their security classes.

References

1. Kalke T (2001) Considerations on watermarking security. In: Proceedings of the of MMSP. Cannes, France, October pp 201–206
2. Hartung F, Su JK, Girod B (1999) Spread spectrum watermarking: malicious attacks and counter-attacks. In: San Jose CA, January S (eds) Proceedings of SPIE: Security and watermarking of multimedia contents Bd. 3657. pp 147–158

3. Florencio M (2003) Improved spread spectrum: a new modulation technique for robust water-marking. In: IEEE Transactions on Signal Processing 51

4. Furon T, Jégourel C, Guyader A, Cérou F (2009) Estimating the probability fo false alarm for a zero-bit watermarking technique. In: Digital Signal Processing, 2009 16th International Conference on IEEE, pp 1–8

5. Miller ML, Cox IJ, Bloom JA (2000) Informed embedding: exploiting image and detector information during watermark insertion. In: Image Processing, 2000. Proceedings. 2000 International Conference on Bd. 3 IEEE, p 1–4

6. Furon T, Bas P (2008) Broken arrows. EURASIP J Inf Secur 1–13:1687–4161

7. Costa M (1983) Writing on dirty paper. In: IEEE Trans. on Inf Theor. 29:439–441

8. Miller ML, Doërr GJ, Cox IJ (2004) Applying informed coding and embedding to design a robust, high capacity watermark. In: IEEE Trans. on Imag Process. 6:791–807

9. Chen B, Wornell GW (2001) Quantization index modulation : a class of provably good methods for digital watermarking and information embedding. IEEE Trans Inf Theor 47(4):1423–1443

10. Eggers JJ, Buml R, Tzschoppe R, Girod B (2003) Scalar Costa Scheme for Information Embedding. In: IEEE Trans Signal Process. 51: 1003–1019

11. Moulin P, Koetter R (2005) Data hiding codes. In: Proceedings of IEEE 93, December 12 pp 2083–2126

12. Conway JH, Sloane NJA (1998) Sphere packings, lattices and groups. Springer, New York

13. Erez U, Litsyn S, Zamir R (2005) Lattices which are good for (almost) everything. IEEE Trans Inf Theor 51:3401–3416

14. Pérez-Freire L, Pérez-González F (2008) Security of lattice-based data hiding against the watermarked only attack. In: IEEE Trans Inf Forensics Secur 3, 4:593–610. http://dx.doi.org/10.1109/TIFS.2008.2002938. doi:10.1109/TIFS.2008.2002938. ISSN 1556–6013

15. Andrew JV (1995) CDMA: Principles of Spread Spectrum Communication. Addison-Wesley, Boston ISBN 0–201–63374–4

Chapter 3
Fundamentals

If one of the first paper on digital watermarking published in 1996 was named "Secure Spread Spectrum Watermarking For Multimedia" [1], the notion of security, i.e. the presence of an adversary, was only considered in 1998 with the scenario of the Oracle attack (see Sect. 1.3.3) proposed by Kalker et al. [2]. This attack and possible counter-attacks was afterward studied by Mansour in 2002 [3], Choubassi in 2005 [4], and Comesaña et. al in 2005 [5, 6] (see also Sect. 4.3.5).

This chapter continues to pave the road of watermarking security which was first formally defined in 2001 by Kalker as *"the inability by unauthorized users to have access to the raw watermarking channel"* [7] in 2001. This notion has been more extensively detailed by Barni et al. in 2003 [8] with the proposal of attack scenarios inspired from the classification of Diffie and Hellman [9].

3.1 Introduction: Information Theoretical Approaches

The secret key is usually a signal: In SS, the secret key is the set of carriers; in Quantization Index Modulation schemes (QIM), it is the dither randomizing the quantization [10]. This signal is usually generated at the embedding and decoding sides thanks to a pseudo-random generator fed by a seed. However, the adversary has no interest in disclosing this seed, because, by analyzing watermarked contents, it is usually simpler to directly estimate $\mathbf{k}$ without knowing this seed.

The application of the information theoretic approach of C.E. Shannon allowed to quantify watermarking security levels [11–13]. This theory regards the signals used at the embedding as random variables (r.v.). Let us denote $\mathbf{K}$ the r.v. associated to the secret key, $\mathcal{K}$ the space of the secret keys, $\mathbf{X}$ the r.v. associated to the host, $\mathcal{X}$ the space of the hosts. Before producing any watermarked content, the designer draws the secret key $\mathbf{k}$ according to a given distribution $p_{\mathbf{K}}$. The adversary knows $\mathcal{K}$ and

© The Author(s) 2016
P. Bas et al., *Watermarking Security*, SpringerBriefs in Signal Processing,
DOI 10.1007/978-981-10-0506-0_3

p_K but he doesn't know the instantiation $\mathbf{k}$. The approach consists in measuring this lack of knowledge by the entropy of the key $H(\mathbf{K}) \triangleq - \oint_K p_K(\mathbf{k}) \log_2 p_K(\mathbf{k})$ (i.e., an integral if $\mathbf{K}$ is a continuous r.v. or a sum if $\mathbf{K}$ is a discrete r.v.).

Now, suppose the adversary sees N_o observations denoted as $\mathbf{O}^{N_o} = \{\mathbf{O}_1, \ldots, \mathbf{O}_{N_o}\}$. The question is whether this key will remain a secret once the adversary gets the observations. These include at least some watermarked contents which have been produced by the same embedder (same algorithm $e(\cdot)$, same secret key $\mathbf{k}$). These are also regarded as r.v. $\mathbf{Y}$. The observations may also encompass some other data depending on the attack setup (see definitions of WOA, KMA, KOA in [12]). Note that this section focuses on the KMA (Known Message Attack) scenario where observations are pairs of a watermarked content and its embedded message: $\mathbf{O}_i = (\mathbf{Y}_i, M_i)$.

By carefully analyzing these observations, the adversary might deduce some information about the secret key. The adversary can refine his knowledge about the key by constructing a posteriori distribution $p_K(\mathbf{k}|\mathbf{O}^{N_o})$. The information leakage is given by the mutual information between the secret key and the observations $I(\mathbf{K}; \mathbf{O}^{N_o})$, and the equivocation $h_e(N_o) \triangleq H(\mathbf{K}|\mathbf{O}^{N_o})$ determines how this leakage decreases the initial lack of information: $h_e(N_o) = H(\mathbf{K}) - I(\mathbf{K}; \mathbf{O}^{N_o})$. The equivocation is always a non increasing function. With this formulation, a perfect covering is tantamount to $I(\mathbf{K}; \mathbf{O}^{N_o}) = 0$. Yet, for most of the watermarking schemes, the information leakage is not null. If identifiability is granted, the equivocation about the secret key decreases down to 0 ($\mathbf{K}$ is a discrete r.v.) or $-\infty$ ($\mathbf{K}$ is a continuous r.v.) as the adversary keeps on observing more data.

This methodology needs p_X, p_K and $e(\cdot)$ to derive the distribution of the observations and, in the end, the equivocation. There is no use of the decoding algorithm. It has been successfully applied to additive SS and ISS schemes [11] for Gaussian distribution p_X and to the lattice-based DC-QIM (Distortion Compensated Quantization Index Modulation) scheme [14] under the flat host assumption (p_X is constant at the scale of the watermark signal).

The goal of this chapter is to present two complementary tools that can be used to classify or measure the security of a watermarking scheme:

1. In Sect. 3.2, we propose different classes of watermarking schemes, each class limiting the possibility of attack that the adversary can do, as we shall see in Chap. 4, it is possible to design embedding schemes belonging to these different classes.
2. In Sect. 3.3 we propose a way to compute the effective key length of a watermarking scheme. This security measure is more practical than information theoretic measures such as the entropy of the key or the equivocation because it takes into account the fact that two different keys can give access to the same hidden information.

3.2 Watermarking Security Classes

Watermarking security was first considered from the point of view of security level assessment. In [12], the Diffie and Hellman methodology is adapted to digital watermarking and yields a classification of the attacks according to the type of information the adversary has access to:

- Known-Message Attack (KMA) occurs when an adversary has access to several pairs of watermarked contents and corresponding hidden messages,
- Known-Original Attack (KOA) occurs when an adversary has access to several pairs of watermarked contents and their corresponding original versions,
- Watermark-Only Attack (WOA) occurs when an adversary has only access to several watermarked contents.

This classification has been further extended with the Constant-Message Attack (CMA) [15] where the adversary observes several watermarked contents and only knows that the unknown hidden message is the same in all contents.

The Kerckhoffs' principle [16] states that Alice and Bob shall only rely on some previously shared secret. It *also* states that Alice and Bob must consider that the adversary knows everything on their communication process but their secret.

3.2.1 Embedding Security Classes

This section aims at looking back at the prisoners' problem assuming that the warden and the prisoners all performed a detailed Kerckhoffs analysis of the way secret information is *embedded* into host contents. In this context, we assume the most general case where the adversary only observes generated contents and is not aware of the embedded messages, i.e. the WOA setup.

It should be noted that practically, the secret key is generated using a seed that initializes a pseudo-random number generator with a given output repetition period (PRNG). Therefore, even if one transforms the output of a PRNG to get Gaussian signals, the set of possible Gaussian signals is related to the number of keys of the PRNG and is therefore countable.

3.2.1.1 Definitions of Embedding Security Classes

Applying Kerckhoffs' principle to the embedding function allows to assume that both Alice and the adversary can build a perfect estimation of different pdfs (especially the pdf of the original contents and the pdf of the watermarked contents). The game of security is then defined taking into consideration the knowledge of:

- $p(\mathbf{X})$, the probabilistic model of N_o host contents. Since we are considering the WOA setup, Alice, Bob and the adversary are able to model the joint distribution of $\mathbf{X}$: $p(\mathbf{X}) = p(\mathbf{x}_0, \dots, \mathbf{x}_{N_o-1})$.
- This hypothesis stems from some sort of a worst-case consideration (from Alice and Bob point of view), where the adversary was able to model the original contents.
- $p(\mathbf{Y})$, the probabilistic model of N_o watermarked contents. Note that each content has been watermarked using a different key. This model can be built by the adversary using his knowledge of the embedding function.
- $p(\mathbf{Y_K})$, the probabilistic model of N_o watermarked contents. In this case, each content has been watermarked using the same unknown key $\mathbf{K}$. This is the model that the adversary can build while observing the collection of watermarked contents without any knowledge on the secret key.
- $p(\mathbf{Y}|\mathbf{K}_i)$, the probabilistic model of N_o watermarked contents. Each content has been watermarked using the same known key $\mathbf{K}_i$. This is the model that the adversary can build while applying the Kerckhoffs' principle, e.g. while embedding random messages into a collection of watermarked contents using his own key $\mathbf{K}_i$.

Since host contents are assumed to be independent, the previous models are the products of marginals, i.e.: $p(\mathbf{X}) = p(\mathbf{x}_0) \cdot p(\mathbf{x}_1) \dots p(\mathbf{x}_{N_o-1})$. The same holds for $p(\mathbf{Y})$, and $p(\mathbf{Y}|\mathbf{K})$. Thus, definitions of embedding security classes in the sequel also holds for the marginals. However, we prefer to use joint probabilities in order to highlight the fact that the pirate can accumulate several contents.

Finally, the adversary's ultimate goal is to estimate the constant $\mathbf{K_e}$ which maximizes the likelihood $p(\mathbf{Y_K}|\mathbf{K_e})$.

Definition 1 ([Insecurity]) An embedding function is **insecure** iff (if and only if):

$$\exists! \ \mathbf{K}_1 \in \mathcal{K}, \ p(\mathbf{Y}|\mathbf{K}_1) = p(\mathbf{Y_K}), \tag{3.2.1}$$

where $\exists!$ denotes the unique existence.

An embedding function is then called insecure if there exists an unique key $\mathbf{K}_1$ whose associated model of watermarked contents with this key $p(\mathbf{Y}|\mathbf{K}_1)$ matches the model of the observations $p(\mathbf{Y_K})$. It implies that the maximum likelihood estimation of the secret key is possible, the worst method being the exhaustive search considering the N_k different keys.

Definition 2 ([Key-security]) An embedding function is **key-secure** iff:

$$\exists \ \mathcal{S_K} \subset \mathcal{K}, \text{card}(\mathcal{S_K}) > 1,$$
$$\forall \ \mathbf{K}_1 \in \mathcal{S_K}, \ p(\mathbf{Y}|\mathbf{K}_1) = p(\mathbf{Y_K}). \tag{3.2.2}$$

We can define $\mathcal{S}_{\mathbf{K}}$ as the invariant subset of the key $\mathbf{K}$. Note that we obviously have $\mathbf{K} \in \mathcal{S}_{\mathbf{K}}$. $\mathcal{S}_{\mathbf{K}}$ represents the set of keys which does not modify the probabilistic model of the observations. When a watermarking scheme is said insecure we can claim that $\mathcal{S}_{\mathbf{K}}$ does not exist. If this subset equals $\mathcal{K}$ then the algorithm is called subspace-secure (see next definition).

Note that even if it is impossible to estimate the secret key $\mathbf{K}$ for key-security, it is possible to estimate the secret subspace $\mathcal{S}_{\mathbf{K}}$ and to reduce the uncertainty of the estimation of the secret key. The security of key-secure embedding schemes relies on the number of possible keys included in $\mathcal{S}_{\mathbf{K}}$ which is $\mathrm{card}(\mathcal{S}_{\mathbf{K}})$. As we will see further in Sect. 4.1.2, Circular Watermarking enables to achieve key-security and the invariant-subspace associated to the key is included in an hypersphere.

Note that Doërr et al. [17] defined the subspace related to a secret key for SS watermarking schemes as the set of all keys belonging to the hyperplane where the collection of watermarked signals $\mathbf{Y}_{\mathbf{K}}$ share the same covariance matrix. We can call such subspace a covariant-subspace. The definition of subspace invariance proposed here is more accurate because the density functions are directly considered and not only their second-order statistics. Nevertheless it is important to have the possibility to estimate either the invariant-subspace or the covariant-subspace for security purposes. If one of these subspaces is known, then it is possible to decrease the robustness of the watermarking scheme regarding AWGN attack for example and to design a random worst case attack where the attacking vector $\mathbf{v}$ belongs to the private subspace.

Key-security consequently means that it is impossible for the adversary to estimate the secret key $\mathbf{K}$ even if it is possible to estimate the subspace $\mathcal{S}_{\mathbf{K}}$.

Definition 3 ([Subspace-security]) An embedding function is **subspace-secure** iff :

$$\forall\, \mathbf{K}_1 \in \mathcal{K}, \; p(\mathbf{Y}|\mathbf{K}_1) = p(\mathbf{Y}_{\mathbf{K}}). \tag{3.2.3}$$

Subspace-security means that even in the case of an exhaustive search, the adversary will not be able to distinguish between the right secret key and any wrong key. Consequently, it will be impossible for the adversary to estimate the invariant-subspace $\mathcal{S}_{\mathbf{K}}$ associated with the secret key $\mathbf{K}$. In other words, the conditional-pdf $p(\mathbf{Y}|\mathbf{K})$ does not depend on the key $\mathbf{K}$ which is equivalent to state that $\mathbf{Y}$ and $\mathbf{K}$ are independent.

Note that subspace-security implies key-security: subspace-security allows to choose any two keys $\mathbf{K}_1$ and $\mathbf{K}_2$ for which $p(\mathbf{Y}|\mathbf{K}_1) = p(\mathbf{Y}|\mathbf{K}_2) = p(\mathbf{Y}_{\mathbf{K}})$ holds and to obtain the property of key-security.

It is also important to point out that, by definition, subspace-security implies no information leakage between the watermarked contents and the key as defined by [13]. This is because subspace-security states that the right key is equivalent to any other (wrong) key: the adversary cannot extract any knowledge from her observations. Consequently:

$$\text{Subspace-security} \Leftrightarrow I(\mathbf{Y}_{\mathbf{K}}, \mathbf{K}) = 0. \tag{3.2.4}$$

Fig. 3.1 Diagram for
embedding security classes

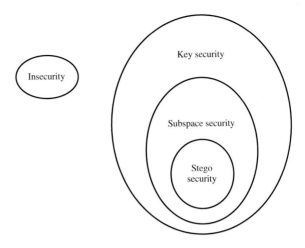

Definition 4 ([Stego-security]) An embedding function is **stego-secure** iff:

$$\forall \mathbf{K}_1 \in \mathcal{K}, \ p(\mathbf{Y}|\mathbf{K}_1) = p(\mathbf{X}). \tag{3.2.5}$$

Stego-security states that knowledge of $\mathbf{K}$ does not help to make the difference
between $p(\mathbf{X})$ and $p(\mathbf{Y})$.

Note that stego-security implies subspace-security. However, subspace-security
does not imply stego-security. This definition implies that $p(\mathbf{Y}|\mathbf{K}_1) = p(\mathbf{Y}|\mathbf{K}_2) = \cdots = p(\mathbf{Y}|\mathbf{K}_{N_k}) = p(\mathbf{Y}) = p(\mathbf{X})$ which is equivalent to a zero Kullback-Leibler
divergence (definition of "perfect secrecy" proposed by Cachin [18]):

$$\text{Stego-security} \Rightarrow D_{KL}(p(\mathbf{Y})||p(\mathbf{X})) = 0. \tag{3.2.6}$$

Practically it says that it is impossible for the adversary to decide whether a content
has been processed through the embedding function or not.

One can finally summarize the relationships between embedding security classes
with the diagram of Fig. 3.1.

3.2.1.2 Adversary's Options

According to which security class the embedding function belongs, the adversary
has several options:

1. if the scheme is stego-secure, the adversary cannot get any information from the
   transmitted contents;
2. if the scheme is subspace-secure but not stego-secure, the adversary is not able
   to estimate $\mathcal{S}_{\mathbf{K}}$ (neither $\mathbf{K}$), but the adversary is able to distinguish stego contents

from innocent ones, e.g. the adversary will be able to perform steganalysis, i.e. to detect the presence of the watermark.

3. if the scheme is key-secure but not subspace-secure, the adversary shall be able to estimate, given enough observations, the subspace $\mathcal{S}_{\mathbf{K}}$ but not the secret key $\mathbf{K}$. The adversary will also be able to concentrate the energy of her attack into the invariant-subspace of the codewords. Practically, this means that it will be possible to jam the message with a smaller distortion than in the previous case.

4. if the scheme is insecure, the estimation of $\mathbf{K}$ is possible and the security of the system is bound to be broken. The adversary will be able to have access to the covert channel. More precisely, in a pure WOA framework, the adversary will also be able only to notice differences between hidden messages or flip the bits while minimizing the distortion (knowledge of some messages is needed to gain full read-write access to the hidden channel).

3.2.1.3   Secure Embeddings

In Chap. 4 we propose different embedding scheme which are either stego secure (Natural Watermarking presented in Sect. 4.1.1 and its informed versions Sect. 4.1.4 or the Soft Scalar Costa Scheme presented in Sect. 4.2.2), or the implementation of Circular Watermarking presented in Sect. 4.1.2.

3.3   Measuring Watermarking Security Using the Effective Key Length

In this section, we propose a new measure of the security of a watermarking scheme called the effective key length.

3.3.1   How to Define the Secret Key in Watermarking?

An early definition of security was coined by Ton Kalker as *the inability by unauthorized users to have access to the raw watermarking channel* [7]. The problem addressed in this section is the following: the methodology to assess the security levels of watermarking schemes, proposed in [11–14, 19], doesn't completely capture T. Kalker's definition. These papers are based on the translation of C.E. Shannon's definition of security for symmetric crypto-systems [20] into watermarking terms. This was the first approach providing important insights on watermarking security. Section 3.1 presents this past approach in more details and outlines the following fact: this methodology only takes into account the embedding side. How could it capture the '*access to raw watermarking channel*' in Kalker's definition if just half of the scheme is considered?

Watermarking and symmetric cryptography strongly disagree in the following point: In symmetric cryptography, the deciphering key is unique and is the ciphering key. Therefore, inferring this key from the observations (here, say some cipher texts) is the main task of the attacker. The disclosure of this key grants the adversary the access to the crypto-channel. This is implicitly assumed in Shannon's framework. Nevertheless watermarking differs from symmetric cryptography by the fact that several keys can reliably decode hidden messages. Therefore, the precise disclosure of the secret key used at the embedding side is one possible way to get access to the watermarking channel, but it may not be the only one.

As a solution, we propose an alternative methodology to assess the security level of a watermarking scheme as detailed in Sect. 3.3.2. It is also inspired by cryptography, but this time, via the brute force attack scenario. In brief, our approach is based on the probability P that the adversary finds a key that grants him the access to the watermarking channel as wished by Kalker: either a key decoding hidden messages embedded with the true secret key, either a key embedding messages that will be decoded with the true secret key. This gives birth to the concept of *equivalent keys* presented in Sect. 3.3.3. Our new definition of the security level is called the *effective key length* and is quantified by $\ell = -\log_2(P)$ in bits. This transposes the notion of cryptographic key length to watermarking: the bigger the effective key length, the smaller the probability of finding an equivalent key. This alternative methodology equally takes into account the embedding and the decoding sides. It is also simpler and more practical because the numerical evaluation of the effective key length is made possible (see Sect. 3.3.5). This is mainly due to the fact that our approach is not based on information theoretical quantities whose closed-form expressions are difficult to derive (if not impossible), and whose estimations by numerical simulation are a daunting task.

3.3.2 The Effective Key Length

In symmetric cryptography, the security is in direct relationship with the length of the secret key, which is a binary word of L bits. The length of the keys L is the entropy in bits if the keys are uniformly distributed but it is also the maximum number of tests in logarithmic scale of the brute force attack which finds the key by scanning the $|K|$ potential keys [21]. The stopping condition has little importance. One often assumes that the adversary tests keys until decoded messages are meaningful. We can also rephrase this with probability: If the adversary draws a key uniformly, the probability to pick the secret key is $P = 2^{-L}$, or in logarithmic scale $-\log_2(P) = L$ bits. With the help of some observations, the goal of the cryptanalysts is to find attacks requiring less operations than the brute force attack. A good cryptosystem has a security close to the length of the key L. For instance, the best attack so far on one version of the Advanced Encryption Standard using 128 bits secret key offers a computational complexity of $2^{126.1}$ [22]. Studying security within a probabilistic framework has also been done in other fields of cryptography (for instance, in authentication [23]).

Our idea is to transpose the notion of key length to watermarking thanks to the brute force attack scenario. A naive strategy is to take the size of the seed of the pseudo-random generator as it is the maximum number of tests of a brute force attack scanning all the seeds. Yet, it doesn't take into account how the secret key is derived from the seed, and especially whether two seeds leads to very similar secret keys. Our approach relies on the probabilistic framework explained below, which takes into account that the secret key may not be unique in some sense.

3.3.2.1 Equivalent Keys

Denote by $\hat{m}$ the message decoded from $\mathbf{y}$ with the secret key $\mathbf{k}$: $\hat{m} = d(\mathbf{y}, \mathbf{k})$. We expect that $\hat{m} = m$, but this might also hold for another decoding key $\mathbf{k}'$. This raises the concept of equivalent keys: for instance, $\mathbf{k}'$ is equivalent to the secret key $\mathbf{k}$ if it grants the decoding of almost all contents watermarked with $\mathbf{k}$ (see mathematical Definitions (3.3.4) and (3.3.5) in Sect. 3.3.3). This idea was first mentioned in [24], where the authors made the first distinction between the key lengths in cryptography and watermarking.

The fact that the decoding key might not be unique creates a big distinction with cryptography. However, the rationale of the brute force attack still holds. The adversary proposes a test key $\mathbf{k}'$ and we assume there is a genie telling him whether $\mathbf{k}'$ is equivalent to $\mathbf{k}$. In other words, the security of the watermarking scheme relies on the rarity of such keys: The lower the probability P of $\mathbf{k}'$ being equivalent to $\mathbf{k}$, the more secure is the scheme. We propose to define the effective key length as a logarithmic measure of this probability. Note that in our proposal, we must pay attention to the decoding algorithm $d(\cdot)$ because it is central to the definition of equivalent keys.

Like in the previous methodology, the attack setup (WOA, KMA, KOA) determines the data from which the test key is derived. In this section, we restrict our attention to the Known Message Attack (KMA—an observation is a pair of a watermarked content and the embedded message: $\mathbf{o}_i = \{\mathbf{y}_i, m_i\}$).

3.3.3 Definition of the Effective Key Length

This section explains the concept of equivalent keys necessary to define the effective key length.

3.3.3.1 Definition of the Equivalent Keys

We define by $\mathcal{D}_m(\mathbf{k}) \subset \mathcal{X}$ the decoding region associated to the message m and for the key $\mathbf{k}$ by:

$$\mathcal{D}_m(\mathbf{k}) \triangleq \{\mathbf{y} \in \mathcal{X} : d(\mathbf{y}, \mathbf{k}) = m\}. \tag{3.3.1}$$

The topology and location of this region in $\mathcal{X}$ depends of the decoding algorithm and of $\mathbf{k}$.

To hide message m, the encoder tries to push the host vector $\mathbf{x}$ deep inside $\mathcal{D}_m(\mathbf{k})$, and this creates an embedding region $\mathcal{E}_m(\mathbf{k}) \subseteq \mathcal{X}$:

$$\mathcal{E}_m(\mathbf{k}) \triangleq \{\mathbf{y} \in \mathcal{X} : \exists \mathbf{x} \in \mathcal{X} \text{ s.t. } \mathbf{y} = e(\mathbf{x}, m, \mathbf{k})\}. \tag{3.3.2}$$

A watermarking scheme provides robustness by embedding in such a way that the watermarked contents are located far away from the boundary inside the decoding region. If the vector extracted from an attacked content $\mathbf{z} = \mathbf{y} + \mathbf{n}$ goes out of $\mathcal{E}_m(\mathbf{k})$, $\mathbf{z}$ might still be in $\mathcal{D}_m(\mathbf{k})$ and the correct message is decoded. For some watermarking schemes (like QIM without distortion compensation), we have $\mathcal{E}_m(\mathbf{k}) \subseteq \mathcal{D}_m(\mathbf{k})$. Therefore, there might exist another key $\mathbf{k}'$ such that $\mathcal{E}_m(\mathbf{k}) \subseteq \mathcal{D}_m(\mathbf{k}')$. A graphical illustration of this phenomenon is depicted on Fig. 3.2. In general even if there is no noise, $\mathcal{E}_m(\mathbf{k}) \not\subset \mathcal{D}_m(\mathbf{k})$. This means that some decoding errors are made even in the noiseless case. We define the Symbol Error Rate (SER) in the noiseless case as $\eta(0) \triangleq \mathbb{P}[d(e(\mathbf{X}, M, \mathbf{k}), \mathbf{k}) \neq M]$. Capital letters $\mathbf{X}$ and M explicit the fact that the probability is over two r.v.: the host and the message to be embedded.

We now define the equivalent keys and the associated equivalent region. We make the distinction between the equivalent decoding keys (the equivalent decoding region) and the equivalent embedding keys (resp. the equivalent embedding region) (Fig. 3.3).

The set of equivalent decoding keys $\mathcal{K}_{eq}^{(d)}(\mathbf{k}, \epsilon) \subset \mathcal{K}$ with $\epsilon \geq 0$ is the set of keys that allows a decoding of the hidden messages embedded with $\mathbf{k}$ with a probability bigger than $1 - \epsilon$:

$$\mathcal{K}_{eq}^{(d)}(\mathbf{k}, \epsilon) = \{\mathbf{k}' \in \mathcal{K} : \mathbb{P}\left[d(e(\mathbf{X}, M, \mathbf{k}), \mathbf{k}') \neq M\right] \leq \epsilon\}. \tag{3.3.3}$$

In the same way, the set of equivalent encoding keys $\mathcal{K}_{eq}^{(e)}(\mathbf{k}, \epsilon) \subset \mathcal{K}$ is the set of keys that allow to embed messages which are reliably decoded with key $\mathbf{k}$:

Fig. 3.2 Graphical representation in $\mathcal{X}$ of three decoding regions. The key $\mathbf{k}'$ belongs the equivalent decoding region $\mathcal{K}_{eq}^{(d)}(\mathbf{k}, 0)$, but not $\mathbf{k}''$

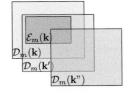

Fig. 3.3 Graphical
representation of $\mathcal{K}$ and the
equivalent region $\mathcal{K}_{eq}(\mathbf{k})$.
The *dotted* boundary
represents the support of the
generative function $g(O^{N_o})$
which is used to draw test
keys when the adversary get
observations

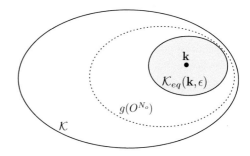

$$\mathcal{K}_{eq}^{(e)}(\mathbf{k}, \epsilon) = \{\mathbf{k}' \in \mathcal{K} : \mathbb{P}\left[d(e(\mathbf{X}, M, \mathbf{k}'), \mathbf{k}) \neq M\right] \leq \epsilon\}. \tag{3.3.4}$$

These sets are not empty for $\epsilon \geq \eta(0)$ since $\mathbf{k}$ is then an element. One expects
that, for a sound design, these sets are empty for $\epsilon < \eta(0)$. Note that for $\epsilon = 0$, these
two definitions are equivalent to:

$$\mathcal{K}_{eq}^{(d)}(\mathbf{k}, 0) = \{\mathbf{k}' \in \mathcal{K} : \mathcal{E}_m(\mathbf{k}) \subseteq \mathcal{D}_m(\mathbf{k}')\}, \tag{3.3.5}$$

and

$$\mathcal{K}_{eq}^{(e)}(\mathbf{k}, 0) = \{\mathbf{k}' \in \mathcal{K} : \mathcal{E}_m(\mathbf{k}') \subseteq \mathcal{D}_m(\mathbf{k})\}. \tag{3.3.6}$$

3.3.3.2 Definition of the Effective Key Length

The effective key length of a watermarking scheme is now explained using these
definitions. For $N_o = 0$, we call $\ell(\epsilon, 0)$ the **basic key length**, i.e. the effective key
length of a watermarking system when no observation is available. The adversary
randomly generates a key and a genie tells him whether this key gives him access
to the watermarking channel. For $\mathbf{K}'$ and $\mathbf{K}$ independent and distributed by $p_{\mathbf{K}}$, the
probability of success is:

$$P^{(d)}(\epsilon, 0) \triangleq \mathbb{E}_{\mathbf{K}}[\mathbb{E}_{\mathbf{K}'}[\mathbf{K}' \in \mathcal{K}_{eq}^{(d)}(\mathbf{K}, \epsilon)]], \tag{3.3.7}$$

where $\mathbb{E}_{\mathbf{K}}[\cdot]$ denotes the expectation over $\mathbf{K}$.

By analogy with the brute force attack in cryptography, the effective key length
translates this probability into bits:

$$\ell^{(d)}(\epsilon, 0) \triangleq -\log_2(P^{(d)}(\epsilon, 0)) \quad \text{bits}, \tag{3.3.8}$$

which is also the logarithm of the average number of guesses needed to find an
equivalent key.

For $N_o > 0$, the adversary can do a better job by inferring some information about $\mathbf{K}$ from the set of observations $\mathbf{O}^{N_o}$. We denote this inferring process by $\mathbf{K}' = g(\mathbf{O}^{N_o})$. The generative function $g(\cdot)$ is either deterministic (e.g., $\mathbf{k}' = \mathbb{E}[\mathbf{K}|\mathbf{O}^{N_o}]$) or stochastic (e.g., $\mathbf{K}' \sim p_{\mathbf{K}|\mathbf{O}^{N_o}}$). The probability of success is as follows:

$$P^{(d)}(\epsilon, N_o) = \mathbb{E}_{\mathbf{K}}[\mathbb{E}_{\mathbf{O}^{N_o}}[\mathbb{E}_{\mathbf{K}'}[\mathbf{K}' \in \mathcal{K}_{eq}^{(d)}(\mathbf{K}, \epsilon)|\mathbf{O}^{N_o}]]]. \tag{3.3.9}$$

Similar definitions are straightforward for $\ell^{(e)}(\epsilon, N_o)$. Note also that for some watermarking schemes like SS (see Sect. 2.2.1) or DC-QIM (see Sect. 2.4), we have $\mathcal{K}_{eq}^{(e)}(\mathbf{k}, \epsilon) = \mathcal{K}_{eq}^{(d)}(\mathbf{k}, \epsilon)$ [25]. For the sake of a simple notation, we use $P(\epsilon, N_o)$ and $\ell(\epsilon, N_o)$ in the sequel.

3.3.3.3 Bounds on the Effective Key Length

Because function $g(.)$ might not be optimal in the sense that it doesn't maximize the probability of success, it only gives an upper bound on the effective key length (provided $\epsilon \geq \eta(0)$):

$$\ell(\epsilon, N_o) \leq -\log_2(P(\epsilon, N_o)) \quad \text{bits.} \tag{3.3.10}$$

This gives the maximum effort needed by the adversary to break the watermarking system. If the bound is small we can conclude that the security is low. However, if the bound is large, we cannot conclude that the security is high. As for cryptanalysis, the security of the system relies on the state of the art of the attacks, represented here by function $g(\cdot)$.

We conclude this section by stating that the value of the effective key length should be clipped to the size of the seed in bits. The rationale is the following. We assume that the pseudo-random generator is public (Kerckhoff's principle) so that nothing prevents the adversary from using this generator. The worst case for him is when any seed, except the true one, produces a key not in the equivalent set. Then a brute force attack on the seed yields an effective key length of the size of the seed in bits:

$$\ell(\epsilon, N_o) \leq L \quad \text{bits,} \tag{3.3.11}$$

which, taking into account (3.3.10), leads to:

$$\ell(\epsilon, N_o) \leq \min\left(L, -\log_2(P(\epsilon, N_o))\right) \quad \text{bits.} \tag{3.3.12}$$

3.3.4 Mathematical Expressions of the Effective Key Length for ISS

The goal of this section is to analyze the trade-off robustness versus security thanks to our new approach applied to one of the most popular class of watermarking schemes: Improved Spread-Spectrum (ISS). We follow the same notations as in Sect. 2.2.1.

3.3.4.1 Equivalent Region and Basic Key Length

Knowing that the secret keys are on the hypersphere, the adversary picks $\mathbf{k}'$ s.t. $\|\mathbf{k}'\| = 1$ and we denote $\mathbf{k}^\top \mathbf{k}' = \cos(\theta)$. Suppose that $m = 0$ is transmitted. Decoding with $\mathbf{k}'$ yields correlation $d' = \mathbf{x}^\top \mathbf{k}' + (\beta - \gamma.\mathbf{x}^\top \mathbf{k}) \cos(\theta)$. From now on, we consider the hyperplane $\mathcal{H} = \mathrm{Span}(\mathbf{k}, \mathbf{k}')$ equipped with the basis $(\mathbf{e}_1, \mathbf{e}_2)$ s.t. $\mathbf{k} = \mathbf{e}_1$ and $\mathbf{k}' = \cos(\theta)\mathbf{e}_1 + \sin(\theta)\mathbf{e}_2$. We denote the projection of $\mathbf{x}$ onto $\mathcal{H}$ by (x_1, x_2). The associated r.v. X_1 and X_2 are i.i.d. and distributed as $\mathcal{N}(0, \sigma_X^2)$. Thus, in the noiseless case, $D' = (1 - \gamma)X_1 \cos(\theta) + X_2 \sin(\theta) + \beta \cos(\theta)$ is Gaussian distributed. A decoding error occurs when $D' < 0$ with probability:

$$\epsilon = \Phi\left(-\frac{\beta \cos\theta}{\sigma_X \sqrt{\sin^2(\theta) + (1 - \gamma)^2 \cos^2(\theta)}}\right) \qquad (3.3.13)$$

For $\eta(0) \leq \epsilon \leq 1/2$, inverting (3.3.13) together with (2.2.6) shows that $\mathbf{k}'$ is an equivalent key if $\theta \leq \theta_\epsilon$ with

$$\cos\theta_\epsilon = \frac{-\Phi^{-1}(\epsilon)10^{\frac{\mathrm{DWR}}{20}}}{\sqrt{N_v + 10^{\frac{\mathrm{DWR}}{10}}(\gamma(2 - \gamma)(\Phi^{-1}(\epsilon))^2 - \gamma^2)}}. \qquad (3.3.14)$$

In words, $\mathcal{K}_{eq}(\mathbf{k}, 0)$ is the hypercap (the intersection of an hypersphere and an hypercone) of axis $\mathbf{k}$ and angle θ_ϵ. $P(\epsilon, 0)$ is the ratio of the solid angles of this hypercap and of the full hypersphere:

$$P(\epsilon, 0) = \left(1 - I_{\cos^2(\theta_\epsilon)}(1/2, (N_v - 1)/2)\right)/2, \qquad (3.3.15)$$

where I is the regularized incomplete beta function.

3.3.4.2 Trade-off Robustness Versus Security for $N_o = 0$

For fixed parameters γ, ϵ and DWR, the basic key length is a decreasing function of N_v (see Fig. 3.8 for SS), contrary to $\eta(\sigma_N)$. This illustrates the trade-off between security and robustness, which is a well known fact for SS in the watermarking security literature [11–13, 19]. The asymptotical value of the basic key length is given by:

$$\lim_{N_v \to \infty} P(\epsilon, 0) = \frac{1}{2} \left(1 - \text{erf} \left(\frac{|\Phi^{-1}(\epsilon)|}{\sqrt{2}} 10^{\frac{\text{DWR}}{20}} \right) \right). \tag{3.3.16}$$

As $N_v \to \infty$, ISS is more robust while its basic key length decreases but does not vanish to 0.

For fixed N_v and DWR, the basic key length ℓ is a decreasing function of $P(\epsilon, 0)$, which in turn is a decreasing function of $\cos^2(\theta_\epsilon)$. Maximizing this latter quantity gives the best security. A simple analysis of the denominator of (3.3.14) shows that, when γ increases from 0, ℓ starts from the basic key length of SS and decreases. The basic key length reaches a minimum at $\gamma = \min(\gamma^{(S)}, \gamma_{\max})$, with

$$\gamma^{(S)} \triangleq (\Phi^{-1}(\epsilon))^2 / (1 + (\Phi^{-1}(\epsilon))^2). \tag{3.3.17}$$

If $\gamma^{(S)} < \gamma_{\max}$, ℓ then increases for $\gamma > \gamma^{(S)}$ and can even be bigger than the basic key length of SS if $\bar{\gamma}^{(S)} < \gamma < \gamma_{\max}$ with $\bar{\gamma}^{(S)} = 2\gamma^{(S)}$. This happens only if $\epsilon > \Phi(-\sqrt{\gamma_{\max}/(2 - \gamma_{\max})})$, which equals 0.159 when $\gamma_{\max} = 1$.

It is interesting to draw the curve giving ℓ as a function of $\eta(\sigma_N)$ when varying γ. Both $\eta(\sigma_N)$ and ℓ decrease as γ starts increasing from 0. Again, this is the trade-off robustness versus security. Yet, as γ keeps on increasing, the situation gets more complex. If $\gamma^{(R)} < \gamma < \gamma^{(S)}$ (resp. $\gamma^{(R)} > \gamma > \gamma^{(S)}$), then this curve turns on the left (resp. right as in Fig. 3.4). We enter here a range of γ where security and robustness can both decrease (resp. increase). It can happen that two different γ gives different key lengths while producing the same robustness. This phenomenon has been discovered for ISS in [11, Fig. 4], but for $N_o = 1$. Our new security approach shows that trading security against robustness does not hold in general even for $N_o = 0$. Figures 3.4 and 3.5 show that a clever tuning of γ renders ISS both more robust and more secure than SS, for $N_o = 0$.

Fig. 3.4 Robustness and security evolution for ISS ($N_v = 180$, DWR = 20 dB, WNR = -20 dB, $\epsilon = 0.2$, $N_o = 0$). ($\star$) computed using the estimation of the equivalent region (Sect. 3.3.5.2) with $N_t = 10^6$

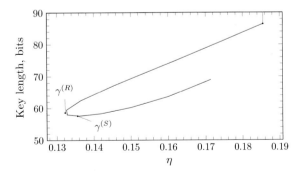

3.3.4.3 Effective Key Length for $N_o > 0$

This subsection proposes some generative functions $\mathbf{K}' = g(\mathbf{O}^{N_o})$ specific to ISS. First, we try deterministic functions taken as estimators of the secret keys under the KMA scenario. Then, we propose a stochastic function build on these estimators.

The pirate observes some watermarked vectors $\mathbf{y}_i$ together with their embedded message m_i. We assume that all messages are equal to 0 without loss of generality (otherwise the pirate works with signals $(-1)^{m_i} \mathbf{y}_i$). According to (2.2.5), we have $\mathbb{E}[\mathbf{Y}_i] = \beta \mathbf{k}$, and a simple estimator is the empirical average:

$$\hat{\mathbf{k}}_{\mathrm{AVE}} = N_o^{-1} \sum_{i=1}^{N_o} \mathbf{y}_i. \tag{3.3.18}$$

This estimator is Gaussian distributed as $\mathcal{N}(\beta \mathbf{k}, \sigma_X^2 N_o^{-1} \mathbf{R})$, with the following covariance matrix:

$$\mathbf{R} = \left(\mathbf{I}_{N_v} + ((1 - \gamma)^2 - 1)\mathbf{k} \cdot \mathbf{k}^\top\right). \tag{3.3.19}$$

In other words, the variance of this estimator is $\sigma_X^2 N_o^{-1}$ in any direction of the space orthogonal to $\mathbf{k}$, along which this variance is $(1 - \gamma)^2$ times smaller. It is also possible to show that with $\mu = \beta$, $\sigma^2 = \sigma_X^2 N_o^{-1}$ and $\sigma_1^2 = (1 - \gamma)^2 \sigma^2$, we have:

$$P(\epsilon, N_o) \approx \left[1 - \mathcal{F}\left(\frac{N_v - 1}{\tan^2(\theta_\epsilon)(1 - \gamma)^2}; 1, N_v - 1, \lambda\right)\right] \Phi\left(\sqrt{\lambda}\right) \tag{3.3.20}$$

with

$$\lambda = \frac{\beta^2 N_o}{\sigma_X^2 (1 - \gamma)^2} = N_o \frac{N_v 10^{-\frac{\mathrm{DWR}}{10}} - \gamma^2}{(1 - \gamma)^2}. \tag{3.3.21}$$

We can also show that the effective key length vanishes to 0 as $N_v \to \infty$ whenever $N_o > 0$, which is a strong difference with the basic key length.

Fig. 3.5 Robustess and security evolution for ISS ($N_v = 180$, **DWR** $= 14\,\mathrm{dB}$, **WNR** $= -7\,\mathrm{dB}$, $\epsilon = 10^{-2}$, $N_o = 0$). ($\star$) computed using the estimation of the equivalent region (Sect. 3.3.5.2) with $N_t = 10^6$

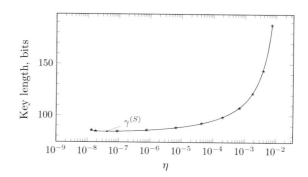

The study of the simple average estimator stems into a better idea: we should also take into account that the direction of minimum variance reveals the true secret key. Indeed, the maximum likelihood estimator $\hat{\mathbf{k}}_{ML} = \arg\max_{\mathbf{k}} \prod_{i=1}^{N_o} p(\mathbf{y}_i|\mathbf{k})$ complies with this idea. Under the Gaussian host assumption and provided $0 < \gamma < 1$, it amounts at minimizing the quantity $\sum_{i=1}^{N_o}(\mathbf{y}_i - \beta\mathbf{k})^\top \mathbf{R}^{-1}(\mathbf{y}_i - \beta\mathbf{k})$. The gradient w.r.t. $\mathbf{k}$ cancels if

$$\mathbf{Y}\mathbf{Y}^\top\hat{\mathbf{k}} = \frac{\beta}{1 - (1-\gamma)^2}\mathbf{Y}\mathbf{1}_{N_o}, \tag{3.3.22}$$

with $\mathbf{Y}$ the $N_v \times N_o$ matrix $(\mathbf{y}_1, \ldots, \mathbf{y}_{No})$ and $\mathbf{1}_{N_o}$ the $N_o \times 1$ vector composed of ones. As noted in [14], $\mathbf{Y}\mathbf{Y}^\top$ has rank N_o (at most) and is therefore not invertible if $N_o < N_v$. Yet, we circumvent this difficulty by restricting the search to the linear estimator $\hat{\mathbf{k}} = \sum_{i=1}^{No} w_i\mathbf{y}_i = \mathbf{Y}\mathbf{w}$ that maximizes the likelihood. The optimum weights are then $\mathbf{w} = \frac{\beta}{1-(1-\gamma)^2}(\mathbf{Y}^\top\mathbf{Y})^{-1}\mathbf{1}_{N_o}$. To end up with an unbiased estimator, we set

$$\hat{\mathbf{k}}_{ML\text{-}LIN} = \frac{\sum_{i=1}^{No} w_i\mathbf{y}_i}{\sum_{i=1}^{No} w_i}. \tag{3.3.23}$$

This brings the property that $\mathbf{k}^\top\hat{\mathbf{K}}_{ML\text{-}LIN} \sim \mathcal{N}(\mu, \sigma_1^2)$ with $\mu = \beta$, $\sigma_1^2 = \sigma^2(1-\gamma)^2$, and $\sigma^2 = \|\mathbf{w}\|^2(\sum_i w_i)^{-2}\sigma_X^2$. For β small, N_o small, and γ close to 1, both estimators generate test keys being more often outside the equivalent region than a purely random test key. This leads to the surprise that these estimators, which are deterministic functions of N_o observations, result in an effective key length bigger than the basic key length (i.e. $N_o = 0$). We solve this paradox in the next section.

3.3.4.4   Forging Stochastic Test Keys

We take advantage here of a phenomenon known as *stochastic resonance* in signal processing [26, 27]. We assume that the strategy of the pirate is to first estimate the secret key, and then to artificially randomize it with noise. The estimator is a deterministic function of the observations $\{\mathbf{y}_i\}_{i=1}^{N_o}$, but the test key is a stochastic process due to the noise addition. We set the test key $\mathbf{K}'$ with the following expression:

$$\mathbf{K}' = a\hat{\mathbf{k}} + b\mathbf{N}, \tag{3.3.24}$$

where $\mathbf{N} \sim \mathcal{N}(\mathbf{0}, \mathbf{I}_{N_v})$ and $a \geq 0$. We aim here at finding a sound couple (a, b) once the estimator $\hat{\mathbf{k}}$ has been computed and $\mathbf{N}$ has been drawn and takes value $\mathbf{n}$. We impose the constraint $\|\mathbf{k}'\| = 1$:

$$a^2\|\hat{\mathbf{k}}\|^2 + b^2\|\mathbf{n}\|^2 + 2a \cdot b\hat{\mathbf{k}}^\top\mathbf{n} = 1. \tag{3.3.25}$$

The tested key is an equivalent key if it lies in the hypercone of axis $\mathbf{k}$ and angle θ_ϵ which amounts to $\mathbf{k}^\top\mathbf{k}' > \cos(\theta_\epsilon)$. If the estimator is such that $\mathbf{k}^\top\hat{\mathbf{k}}$ is distributed as

$\mathcal{N}(\mu, \sigma_1^2)$ (which is the case with the estimators described above) and $\mathbf{N}$ is a white Gaussian noise of unitary variance, then the probability of finding an equivalent key is given by:

$$P = \Phi\left(\frac{a\mu - \cos(\theta_\epsilon)}{\sqrt{a^2\sigma_1^2 + b^2}}\right).$$
(3.3.26)

The justification of this approach based on stochastic resonance lies in the fact that the couple $(a, b) = (\|\hat{\mathbf{k}}\|^{-1}, 0)$ can produce a negative ratio for small μ appearing in (3.3.26): the correlation with $\mathbf{k}$ lies in expectation below the threshold $\cos(\theta_\epsilon)$ yielding to the above-mentioned paradox. In this case, increasing the variance of the projection by the noise addition increases the probability. The same paradox occurs if variance σ_1 is too big (because N_o is small or the estimator is badly designed). In extreme cases, it is better to discard the inaccurate estimator (i.e., $a \approx 0$) and set $b \approx \|n\|^{-1}$. In other words, we are back to the $N_o = 0$ setup. In practice, the pirate finds $(a^\star, b^\star)$ that maximizes the ratio appearing in (3.3.26) under the constraint (3.3.25) thanks to a constrained optimization solver.

3.3.5 Practical Effective Key Length Computations

This section details different experimental protocols to numerically evaluate the effective key length. We first propose a general framework with a high complexity. For correlation-based decoders (such as those of SS, ISS, and CASS), some simplifications occur and stem into a more practical experimental setup. The last subsection shows how to further reduce the complexity with the help of a rare event probability estimator.

3.3.5.1 The General Framework

If we are not limited in term of computational power, the probability $P(\epsilon, N_o)$ can be approximated using a classical Monte-Carlo method. We first generate a set of N_1 random secret keys $\{\mathbf{k}_i\}_{i=1}^{N_1}$. For each of them, we also generate N_2 test keys $\{\mathbf{k}'_{i,j}\}_{j=1}^{N_2}$ computed using N_2 distinct sets of N_o observations. Then, an estimation is:

$$\hat{P}^{(d)}(\epsilon, N_o) = \frac{1}{N_1 N_2} \sum_{i=1}^{N_1} \sum_{j=1}^{N_2} u^{(d)}(\mathbf{k}'_{i,j}, \epsilon),$$
(3.3.27)

where

$$u^{(d)}(\mathbf{k}'_{i,j}, \epsilon) = \begin{cases} 1, & \text{if } \mathbf{k}'_{i,j} \in \mathcal{K}_{eq}^{(d)}(\mathbf{k}_i, \epsilon) \\ 0, & \text{otherwise.} \end{cases}$$
(3.3.28)

The probability $P^{(e)}(\epsilon, N_o)$ is respectively approximated using the indicator function $u^{(e)}(\cdot)$ of $\mathcal{K}^{(e)}$.

For $N_o = 0$, each test key $\mathbf{k}'_{i,j}$ is independently drawn according to $p_{\mathbf{K}}$. For $N_o > 0$ and a given i, we first generate N_2 sets of N_o observations $\mathbf{O}_j^{N_o}$, $1 \leq j \leq N_2$, depending on $\mathbf{k}_i$, and we resort to a specific process of constructing $\mathbf{k}'_{i,j} = g(\mathbf{O}_j^{N_o})$ (see Sect. 3.3.3). Secondly, the equivalent region may not have a defined indicator function. In this case, we generate N_t other contents $\{\mathbf{y}_{i,l}\}_{l=1}^{N_t}$ watermarked with $\mathbf{k}_i$ and the test is satisfied if at least $(1 - \epsilon)N_t$ contents are correctly decoded using $\mathbf{k}'_{i,j}$. Mathematically, for the decoding equivalence:

$$\mathbf{k}'_{i,j} \in \mathcal{K}_{eq}^{(d)}(\mathbf{k}_i, \epsilon) \stackrel{\approx}{\Longleftrightarrow} |\{\mathbf{y}_{i,l} \in \mathcal{D}_{m_i}(\mathbf{k}'_{i,j})\}_{l=1}^{N_t}| > (1 - \epsilon)N_t \qquad (3.3.29)$$

These N_t other watermarked contents play a different role than the N_o watermarked contents used to produce test keys. In this experimental protocol, an estimation of $P^{(d)}(\epsilon, N_o)$ needs $N_1(N_2N_o + N_t)$ embeddings and $N_1N_2N_t$ decodings. Due to the limitation of the Monte-Carlo method, N_1N_2 should be in the order of $1/P^{(d)}(\epsilon, N_o)$ for having a meaningful relative variance of the estimation. The parameter N_t should also be quite big for having a good approximation of the indicator function of $\mathcal{K}_{eq}^{(d)}(\mathbf{k}_i, \epsilon)$. It is reasonable to take $N_t = O(c^{N_v})$ for some constant c where N_v is the dimension of $\mathcal{X}$.

This procedure is generic and it blindly resorts to the embedding and the decoding as black boxes. If we have some knowledge about the watermarking technique, some tricks reduce the complexity of the estimation. First, the probability of finding an equivalent key might not depend on $\mathbf{k}_i$, so that we can restrict to $N_1 = 1$ original key. This is the case for the watermarking techniques studied here.

The keystone of our approach is to base the security level on the evaluation of a probability rather than an information theoretic quantity. Nevertheless, the probability to be estimated might be very weak and out of reach of the Monte-Carlo method. Rare event probability estimators such as [28] are more efficient w.r.t. the runtime. Last but not least, if the equivalent set $\mathcal{K}_{eq}^{(d)}(\mathbf{k}, \epsilon)$ is a region described by few parameters, one can directly estimate the parameters instead of using (3.3.27). The following subsections put into practice these simplifications.

3.3.5.2 Approximation of the Equivalent Region $\mathcal{K}_{eq}^{(d)}$ for Correlation-Based Decoding

The equivalent region $\mathcal{K}_{eq}^{(d)}$ depends on the embedding and decoding. For the additive SS, both processes are so simple that we were able to derive closed-form formula of the probability in Sect. 3.3.4. We suppose now that the embedding is more complex which prevents theoretical derivations whereas the decoding remains correlation-based. In Sect. 3.3.6, CASS [29] plays the role of such an embedding.

For a given host $\mathbf{x}$, we can always express the result of the embedding as

$$\mathbf{y} = e(\mathbf{x}, m, \mathbf{k}) = a(\mathbf{x}, m)\mathbf{k} + b(\mathbf{x}, m)\mathbf{u}_\perp(\mathbf{x}, m), \tag{3.3.30}$$

where $\mathbf{k}^\top \mathbf{u}_\perp(\mathbf{x}, m) = 0$. The decoding with $\mathbf{k}'$ is based on the quantity:

$$\mathbf{y}^\top \mathbf{k}' = a(\mathbf{x}, m)\cos(\theta) + b(\mathbf{x}, m).(\mathbf{k}'^\top \mathbf{u}_\perp(\mathbf{x}, m)), \tag{3.3.31}$$

whose sign yields the decoded bit $\hat{m}$. It is important to note that the decoding using $\mathbf{k}'$ can be studied in the 2 dimensional space spanned by $(\mathbf{k}, \mathbf{u}_\perp(\mathbf{x}, m))$. The Symbol Error Rate is expressed in term of the CDF of the statistical r.v. $\mathbf{Y}^\top \mathbf{k}'$ which depends on θ, and is thus denoted $\mathsf{SER}(\theta)$. For $\theta = 0$, we have $\mathsf{SER}(0) = \eta(0)$. For $\epsilon \geq \eta(0)$, we define

$$\theta_\epsilon = \max_{\mathsf{SER}(\theta)=\epsilon} \theta. \tag{3.3.32}$$

This shows that the equivalent decoding region is a hypercone of axis $\mathbf{k}$ and angle θ_ϵ, which depends on the embedding. The only thing we need is to experimentally estimate angle θ_ϵ. Then, Eq. (3.3.15) provides an approximation of the effective key length.

The estimation of θ_ϵ is made under the following rationale. A vector $\mathbf{y}$ watermarked by $\mathbf{k}$ with $m = 0$ is correctly decoded by any $\mathbf{k}'$ s.t. $\mathbf{k}'^\top \mathbf{k} \geq \cos(\theta_\epsilon)$ if its angle ϕ with $\mathbf{k}$ is such that $\phi \in [\theta_\epsilon - \pi/2, \theta_\epsilon + \pi/2]$ (see Fig. 3.6). In practice, we generate N_t contents $\{\mathbf{y}_i\}_{i=1}^{N_t}$ watermarked with $m = 0$, and we compute their angles $\{\phi_i\}_{i=1}^{N_t}$ with $\mathbf{k}$. Once sorted in increasing order, we iteratively find the angle $\phi_{\min}$ such that $\mathrm{int}((1 - \epsilon)N_t)$ vectors have their angle $\phi \in [\phi_{\min} - \pi/2, \phi_{\min} + \pi/2]$ and set $\hat{\theta}_\epsilon = \pi/2 - \phi_{\min}$.

A much lower number N_t of watermarked vectors is needed to accurately estimate one parameter than for a full region of the space. N_t and N_v directly impact the accuracy of $\hat{\theta}_\epsilon$, but since this boils down to the estimation of a single parameter, the order of magnitude of N_t is rather low in comparison with the effective key length.

Fig. 3.6 Projections of $N_t = 5000$ watermarked vectors ($N_v = 60$, $m = 1$) on $\mathbf{k}$ and $\mathbf{u}_\perp$, DWR $= 10\,\mathrm{dB}$, $N_v = 60$, $\epsilon = 10^{-2}$. The vector $\mathbf{k}'_{\max}$ correctly decodes $[(1 - \epsilon)N_t]$ contents

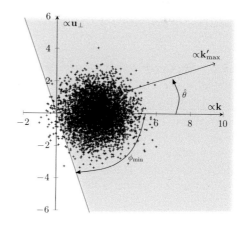

For example, at $N_v = 60$ and DWR $= 10\,$dB, we generate $N_t = 10^6$ contents in order to obtain a reliable effective key length of more than 100 bits. Moreover, the angle θ_ϵ is the same for any $\mathbf{k}$ and the estimation is to be done only once. This avoids counting correct decodings over N_t vectors by (3.3.29).

3.3.5.3 Rare Event Probability Estimator

A fast rare event probability estimator is described in [30]. We explain its application for the correlation-based decoder here, but we also used it for the DC-QIM watermarking scheme in [25]. This algorithm estimates the probability $\mathbb{P}\left[s(\mathbf{K}') \le 0\right]$ under $\mathbf{K}' \sim p_{\mathbf{K}'}$. It needs two ingredients: the generation of test keys distributed according to $p_{\mathbf{K}'}$ and a soft score function $s(\cdot) : \mathcal{K} \to \mathbb{R}$.

For $N_o = 0$, $p_{\mathbf{K}'}$ is indeed $p_{\mathbf{K}}$ from which one can easily sample. In our simulation, we work with an auxiliary random vector $\mathbf{W} \sim \mathcal{N}(\mathbf{0}, \mathbf{I}_{N_v})$. The generator draws $\mathbf{W}$ and outputs a test key $\mathbf{K}' = \mathbf{W}/\|\mathbf{W}\|$. Since the distribution of $\mathbf{W}$ is isotropic, $\mathbf{K}'$ is uniformly distributed over the hypersphere. For $N_o > 0$, $p_{\mathbf{K}'}$ may be unknown. To generate a test key, we first sample N_o independent host contents distributed as $p_{\mathbf{X}}$, we watermark them with $\mathbf{k}$, and finally apply function $g(.)$. The algorithm draws n such test keys, and iteratively modifies those having a low score. The properties of this algorithm depends on n as given in [30]. Qualitatively, the bigger n is, the more accurate but slower is this estimator.

If $\mathcal{K}_{eq}^{(d)}$ is known (Sect. 3.3.4) or approximated (Sect. 3.3.5.2), the score function is simply a 'distance' between the test key and the equivalent region, this distance being zero if the test key is inside. For the schemes analyzed in this section, we work with $s(\mathbf{k}') = |\cos(\theta_\epsilon) - \mathbf{k}'^\top\mathbf{k}|^+$, with $|x|^+ = x$ if x is positive, and 0 otherwise. If $\mathcal{K}_{eq}^{(d)}$ is not known, we generate N_t contents $\{\mathbf{y}_i\}_{i=1}^{N_t}$ watermarked with $\mathbf{k}$, and the score function is the int(ϵN_t)-th smallest 'distance' from these vectors to the set $\mathcal{D}_m(\mathbf{k}')$, where int(.) denotes the closest integer function. In the end, the algorithm returns an estimation of the probability that int$((1-\epsilon)N_t)$ of these vectors are correctly decoded when $\mathbf{K}'$ is distributed as $p_{\mathbf{K}'}$.

3.3.6 Security Analysis of Watermarking Schemes

The goal of the experimental part is threefold. First, we wish to assess the soundness of the experimental measurement presented in Sect. 3.3.5 with a comparison to the theoretical results for SS and ISS. Secondly, we aim at analyzing the impact of both the embedding parameters N_v and DWR and the security parameters ϵ and N_o on the effective key length. Third, we would like to illustrate the interplay between security and robustness for both SS and ISS.

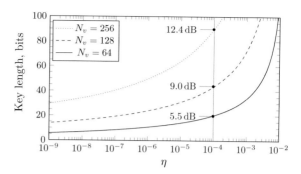

Fig. 3.7 Security ($\epsilon = 10^{-2}$) versus robustness (**WNR** = 0 dB) for SS ($\gamma = 0$). The plot is computed by varying **DWR**, the ticks show the values of **DWR** for $\eta = 10^{-4}$

3.3.6.1 SS and ISS

Figure 3.7 illustrates the trade-off between security and robustness for SS plotting $\ell(\epsilon, 0)$ as a function of $\eta(\sigma_N)$: for a given N_v, a bigger watermarking *power* implies a bigger robustness but a lower security. However, for a constant watermark *energy* D, the robustness is fixed ($\gamma = 0$) while the basic key length increases with N_v.

Figure 3.8 shows the evolution of the basic key length w.r.t. N_v for a constant watermark *power* (i.e. fixed **DWR**). Contrary to a statement of [24, Sect. 4.1], the effective key length is not proportional to N_v. We also note the relatively fast convergence to the strictly positive asymptote (3.3.16), especially at high embedding distortions.

Figure 3.9 highlights the decrease of this asymptotic effective key length with the watermark power. The basic key length at $\epsilon = 0.01$ is computationally significant, say above 64 bits, only for a **DWR** greater than 12 dB. We also rediscover the classical sign ambiguity of some security attacks [12] when the adversary ignores the embedded bit: $\lim_{\mathbf{DWR} \to -\infty} \ell = 1$ bit (see Fig. 3.9 and (3.3.16)). The probability of drawing a test key in the adequate half-space tends to 0.5 when the watermarking power is super strong. At last, Fig. 3.9 shows the decrease of the basic key length with ϵ: the more stringent the access to the watermarking channel, the higher the security is.

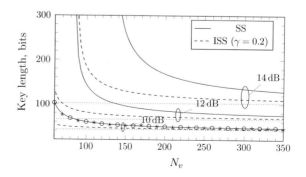

Fig. 3.8 The basic key lengths for $\epsilon = 10^{-2}$ and **DWR** $\in \{10, 12, 14\}$. (*plain lines*) theoretical expression (3.3.15), (○) estimation of the equivalent region (Sect. 3.3.5.2) with $N_t = 10^6$, (⋆) rare event analysis (Sect. 3.3.5.3) with $N_t = 5.10^4$ and $n = 80$, (*dotted lines*) the asymptotes (3.3.16)

Fig. 3.9 Basic key length
for hosts of infinite length
given in (3.3.16)

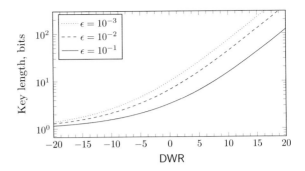

Impact of N_o

We now evaluate the impact of the number of observations in the KMA setup for
SS and ISS. For ISS, Fig. 3.10 illustrates the dramatical collapse of the effective key
length when observations are available. For example, at $\mathsf{DWR} = 10\,\mathrm{dB}$, $N_v = 300$
and $\epsilon = 10^{-2}$, the effective key length drops from roughly 50 bits to nearly 0
bits within 10 observations. Note also that the approximation (3.3.20) is very close
to the Monte Carlo estimations. Figure 3.11 illustrates the paradox highlighted in
Sect. 3.3.3.3: at $\mathsf{DWR} = 12\,\mathrm{dB}$, $N_v = 256$ and $\epsilon = 10^{-2}$, both the average estimator
given by (3.3.18) and the linear estimator (3.3.23) provide upper bounds of the
effective key length for $N_o \in \{1, 2\}$ that are above the basic key-length. A classical
estimator used so far in watermarking security [14] can lead to a security attack less
powerful than a random guess: for $N_o = 1$ the effective key length obtained using the
Average estimator is nearly 3 times larger than the effective key length obtained for
$N_o = 0$! On the other hand, the use of a stochastic estimator minimizing (3.3.26) (a
numerical solver is needed in this case) yields to what one expects, i.e. a decreasing
effective key length w.r.t. N_o. Note also that the key lengths obtained for the stochastic
average estimator and the stochastic ML-Lin average estimator are roughly the same.

The practical methods (Monte-Carlo, rare-event estimator or equivalent region
estimation) match the literal formula (3.3.15) and (3.3.20) either for small or
large effective key lengths on Figs. 3.8, 3.10 and 3.11. The rare event estimator

Fig. 3.10 Effective key
lengths for $\epsilon = 10^{-2}$,
$N_o = 10$, and different DWR
using approximation (3.3.20)
and Monte-Carlo simulations
of Sect. 3.3.5.1 ($\diamond$) with
$N_1 = 1$ and $N_2 = 10^6$

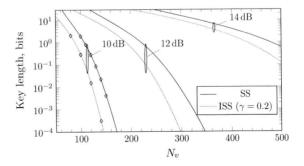

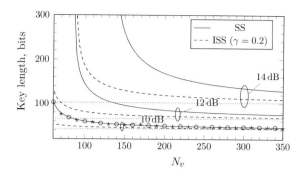

Fig. 3.11 Evolution of the effective key lengths for ISS using $\gamma = 0.2$ w.r.t. different estimators. "RE" stands for Rare Event probability estimator, "Stoc." the stochastic estimator, "Ave." the average estimator and "ML-Lin" the estimator based of the linear maximum likelihood estimator. $N_v = 256$, DWR $= 12$ dB and $\epsilon = 10^{-2}$

(Sect. 3.3.5.3) and the estimator based on $\hat{\theta}_\epsilon$ (Sect. 3.3.5.2) are particularly accurate for large key lengths (see Fig. 3.8), whereas the Monte-Carlo estimator is more efficient for small key length (see Fig. 3.10).

Note also that it is possible to download the Matlab or R source codes that have been used to draw the different plots of this section at at http://people.rennes.inria. fr/Teddy.Furon/weckletoolbox.zip.

3.3.6.2 Zero-Bit Watermarking

To illustrate how it is practically possible to compute an estimation of the key length, we analyze two similar zero-bit embedding methods proposed by Comesaña et al. [31] and Furon and Bas [32]. Both schemes use the normalized correlation as a detection function and the detection function is given by:

$$d(\mathbf{y}, \mathbf{k}) = 1, \quad \text{if} \quad \frac{|<\mathbf{y}, \mathbf{k}>|}{|\mathbf{y}|.|\mathbf{k}|} \geq \cos \alpha,$$
$$d(\mathbf{y}, \mathbf{k}) = 0 \text{ else.} \tag{3.3.33}$$

The angle α is computed according to the probability of false-alarm $p_{fa} = \mathbb{P}[d(\mathbf{X}, \mathbf{k}) = 1)]$. The decoding region for these two schemes is a double hypercone of axis $\mathbf{k}$ and angle α. Without lost of generality, we set $|\mathbf{k}| = 1$ and the set of all possible keys $\mathcal{K}$ is consequently represented by an unitary hypersphere of dimension N_v.

The two embeddings consist in moving the host vector $\mathbf{x}$ into the closest cone by pushing it by a distance D. Using information theoretic arguments [31] propose the OBPA embedding (for Orthogonal to the Boundary and Parallel to the Axis) which first pushes $\mathbf{x}$ in a direction orthogonal to the cone boundary, and then moves afterwards the content parallel to the cone axis. This is proven to maximize asymptotically

Fig. 3.12 Embeddings
proposed by Comesaña et al.
(OBPA) and Furon and Bas
(BA)

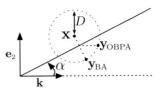

the robustness regarding the AWGN channel. The embedding proposed in [32] called BA embedding (for Broken Arrows, the name of the watermarking system), uses worst case attack arguments to first push $\mathbf{x}$ in a direction orthogonal to the cone boundary and continue in a direction orthogonal to the boundary. If the cone axis is reached however, it goes parallel to the axis. Represented in a plan $\mathcal{P} = (O, \mathbf{k}, \mathbf{e}_2)$ with $\mathbf{e}_2$ a vector such that $\mathbf{x} = x_1\mathbf{k} + x_2\mathbf{e}_2$ and $\mathbf{y} = y_1\mathbf{k} + y_2\mathbf{e}_2$, we can illustrate using Fig. 3.12 the geometrical representations of these two embedding strategies in $\mathcal{P}$. Note that the main difference between these two embeddings is the fact that for a given distortion D, the watermarked contents will be closer to the cone axis using the BA embedding than using the OBPA embedding.

No Observation ($N_o = 0$)
Contrary to watermarking schemes proposed in [33], it is not possible to derive a literal expression of (3.3.35) and we want here to infer an approximation of the equivalent region $\mathcal{K}_{eq}(\mathbf{k}, \epsilon)$ from a set of watermarked contents. Our goal is consequently to compute the maximum possible deviation $\mathbf{k}'$ of the secret key $\mathbf{k}$, such that at least a ratio $(1 - \epsilon)$ of the watermarked contents is included into the hyper-cone of axis $\mathbf{k}'$ and angle α. Let us denote by θ the angle between $\mathbf{k}'$ and $\mathbf{k}$. The equivalent region $\mathcal{K}_{eq}(\mathbf{k}, \epsilon)$ is consequently the union of two spherical caps which is the intersection of the double hyper-cone of axis $\mathbf{k}$ and solid angle θ and the unitary N_v-D hypersphere. $P(\epsilon, 0)$ corresponds to the ratio between the surface of one spherical cap of solid angle θ and the surface of half the sphere (see Eq. (8) of [34]).

$$P_{NC}(\epsilon, 0) = 1 - I_{\cos^2 \theta}(1/2, (N_v - 1)/2), \qquad (3.3.34)$$

where NC stands for a detection using Normalized Correlation. Applying (3.3.35), the key length is given by

$$\ell_{NC}(\epsilon, 0) = -\log_2 \left(1 - I_{\cos^2 \theta}(1/2, (N_v - 1)/2)\right), \qquad (3.3.35)$$

where $I_x(a, b)$ is the regularized incomplete beta function.
   Our problem now consists in finding $\hat{\theta}$ such that:

$$\hat{\theta} = \max\{\theta : \mathbb{P}(d(\mathbf{y}, \mathbf{K}') = 1) = \epsilon, \mathbf{k}^t\mathbf{k}' = \cos\theta\}. \qquad (3.3.36)$$

This can be estimated in practice using a set of N_c watermarked contents included in $\mathcal{D}(\mathbf{k})$:

Fig. 3.13 Geometry when
the detection region is a
hyper-cone

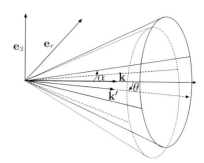

$$\hat{\theta} = \max \left\{ \theta : |\{ \mathbf{y}_i : \mathbf{y}_i \in \mathcal{D}(\mathbf{k}') \}| = [(1 - \epsilon) N_c] \right., \tag{3.3.37}$$

$$\text{and } \mathbf{k}^t \mathbf{k}' = \cos \theta \right\}, \tag{3.3.38}$$

where $1 \leq i \leq N_c$ and $[.]$ denotes the nearest integer function.

It is possible to perform this estimation in a 3D space instead of a N_v-D space by picking a random unitary basis vector $\mathbf{e}_r$, orthogonal to $\mathbf{k}$, and computing the rotation of $\mathbf{k}$ in the plane $(\mathbf{k}, \mathbf{e}_r)$. The test $\mathbf{y} \in \mathcal{D}(\mathbf{k}')$ is then performed in two steps:

1. each content $\mathbf{y}$ is projected onto the orthonormal basis $(\mathbf{k}, \mathbf{e}_r, \mathbf{e}_3)$ where $\mathbf{e}_3$ is such that $\mathbf{y} = y_1 \mathbf{k} + y_2 b_i \mathbf{e}_r + y_3 \mathbf{e}_3$. Note that it is still possible to perform the test $\mathbf{y}_i \in \mathcal{D}(\mathbf{k}')$ using this particular projection.
2. the coordinates of $\mathbf{y}$ in the basis $(\mathbf{k}', \mathbf{e}'_r, \mathbf{e}_3)$, i.e. the basis related to the cone of axis $\mathbf{k}'$, are computed by $\mathbf{k}' = (\cos \theta, \sin \theta, 0)$ and $\mathbf{e}'_r = (-\sin \theta, \cos \theta, 0)$.

A geometric illustration of the two cones in the 3D space is illustrated on Fig. 3.13. The test $\mathbf{y} \in \mathcal{D}(\mathbf{k}')$ is then equivalent to:

$$\frac{y'_1}{\sqrt{y_1^2 + y_2^2 + y_3^2}} \geq \cos \alpha \tag{3.3.39}$$

with $y'_1 = y_1 \cos \theta + y_1 \sin \theta$. The search of $\hat{\theta}$ satisfying (3.3.37) can be done iteratively using a dichotomic search because the number of contents satisfying (3.3.39) is a decreasing function w.r.t. θ. Note that in order to increase the accuracy of $\hat{\theta}$, we can sequentially draw several vectors $\mathbf{e}_r$ and average the results of each estimation.

Figure 3.14 shows the evolution of the key length w.r.t. the DWR (Document to Watermark power Ratio) for the two embeddings with $N_v = 128$, $\epsilon = 0.05$, and $p_{fa} = 10^{-4}$. The key lengths are computed using (3.3.35) and Monte-Carlo simulations with rare event analysis [34] on 1000 watermarked vectors. As expected the key length grows according to the DWR and can reach sizes over 100 bits for $DWR > 6\,\text{dB}$. Notice that the key length of the BA embedding is smaller than the one of OBPA. This is due to the fact that, with BA, the watermarked contents are closer to the hyper-cone axis and consequently the size of the equivalent region is bigger than the one of OBPA. For BA, when the embedding distortion is very important

Fig. 3.14 Key-length
evolution according to the
embedding distortion using
the proposed approximation
and Rare Event estimation
(RE) for $N_v = 128$, $\epsilon = 0.05$
and $p_{fa} = 10^{-4}$

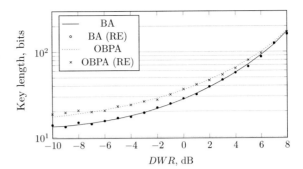

$(DWR \to -\infty)$ all the contents tend to be located on the cone axis which means
that $\theta \to \alpha$ and $\ell_{NC}(0, 0) \to -\log_2 p_{fa}$. The gap between the two key lengths
decreases w.r.t. the embedding distortion because both embedding tends to behave
the same way for small distortion since the first step, moving toward the boundary,
is identical.

Note also that robustness and security are not antagonist here: OBPA, the most
robust scheme w.r.t. the AWGN channel, provides also the longest effective key
length.

$N_o \neq 0$

In this setup $O^{N_o} = Y^{N_o} = (Y_1, Y_2, \ldots, Y_{N_o})$ and we propose to use the principal component of the observations Y^{N_o}, i.e. the eigenvector associated to the most
important eigenvalue of the covariance matrix $\mathbf{C_Y} = N_o^{-1} Y^{N_o} (Y^{N_o})^t$ as a guessing
key $\mathbf{k}'$. A similar strategy was previously used to evaluate the security of Broken-
Arrows during the BOWS-2 challenge [35]. If $N_o < N_v$, we can compute the Eigen
decomposition of the Gram matrix $\mathbf{G_Y} = (Y^{N_o})^t Y^{N_o}$ instead (see [36], Sect. 12.1.4).

Figure 3.15 presents the evolution of the key size in the same setup than in the
previous subsection ($N_v = 128$, $\epsilon = 0.05$, $p_{fa} = 10^{-4}$) for two embedding distortions ($DWR = 5\,\mathrm{dB}$ and $DWR = 7\,\mathrm{dB}$). Monte-Carlo simulation using 10^8 sets of
N_o contents where used in this experiment combined with the extraction of principal
components. We can observe the tremendous reduction of the key size for these two

Fig. 3.15 Key length
evolution according to N_o
($N_v = 128$, $\epsilon = 0.05$,
$p_{fa} = 10^{-4}$) for $DWR = 5$
dB and $DWR = 7$ dB

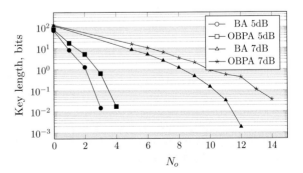

schemes when watermarked contents are available to the attacker. The key length of BA decreases faster than the key length of OBPA. This is due to the fact that variance of the contents along directions orthogonal to **k** is smaller for BA than OBPA and this favors an accurate estimation of the most principal component.

3.3.6.3 The Practical Example of the BOWS-2 Contest

This subsection applies our methodology to approximate the effective key length of the BA scheme of the BOWS-2 international contest [37]. This helps understanding if it was possible to find a key by random guess. We recall that the host vector was extracted from a 512×512 image and $N_v = 258, 048$. In a subspace of dimension 256, the detection region was built as the union of 30 double-hyper-cones with orthogonal axis. The secret key was consequently defined by the basis vectors of the 256 dimensional subspace. By assuming that the subspace is public (but not its basis vectors), we can compute an equivalent region which is larger than the real one and consequently find an upper bound of $P_{BOWS}(\epsilon, 0)$ and a lower bound of $\ell_{BOWS}(\epsilon, 0)$. $P_{BOWS}(\epsilon, 0)$ is upper bounded by the probability of drawing 30 orthogonal vectors falling each in a different equivalent region associated to a true axis:

$$P_{BOWS}(\epsilon, 0) < \Pi_{i=1}^{30} P_{NC}(\epsilon, 0, N_v = 256 - i + 1)$$
$$< P_{NC}(\epsilon, 0, N_v = 226)^{30}, \quad (3.3.40)$$

with the parameter $N_v = 256 - i + 1$ coming from the fact that the vector axes are orthogonal. Hence:

$$\ell_{BOWS}(\epsilon, 0) > 30.\ell_{NC}(\epsilon, 0, N_v = 226). \quad (3.3.41)$$

In practice, we compute $\ell_{NC}(\epsilon, 0, N_v = 226)$ using (3.3.35) on 10,000 images watermarked using the same embedding setup than during the contest ($PSNR = 43$ dB, $p_{fa} = 3.10^{-6}$). We obtain $\ell_{NC}(0.05, 0, N_v = 226) \approx 35$ bits and $\ell_{BOWS}(0.05, 0) > 1050$ bits. On the other hand, since the pseudo random-generator used a 128 bits long seed within the C implementation of the algorithm, this length is an implicit upper bound of the true key length and we can finally conclude that $\ell_{BOWS}(0.05, 0) = 128$ bits. This confirms the idea that the random exhaustive search was impossible during the contest.

3.4 Conclusions of the Chapter

This chapter has answered two main questions in watermarking security:

1. How to measure the security of a given scheme using tools such as the effective key length or the mutual information.

2. What are the different security classes of a watermarking scheme? Each security classes defining the set of possible attacks offered to the adversary.

However, the reader has to be aware that security, distortion and robustness are intertwined constraints and that if one watermarking scheme can be claimed to be secure, or say have a secret key of a given length, other questions still need to be answered. Among them we can list:

1. What is the secure capacity of a watermarking scheme? (how much information can be conveyed in a secure way?)
2. What are the optimal key estimator? i.e. the estimator that minimize the key length? Regarding this question, the answer is probably the same than in Cryptography, the best attack have yet to be designed.
3. How to minimize the distortion of a secure scheme? If the 2 first problems remains still open, the next chapter answers partly to the last one.

References

1. Cox I, Killian J, Leighton T, Shamoon T (1996) Secure spread spectrum watermarking for images, audio and video. In: Proceedings of the IEEE international conference on image processing ICIP-96, Lausanne, Switzerland, pp 243–246
2. Kalker T, Linnartz JP, Dijk van M (1998) Watermark estimation through detector analysis. In: IEEE international conference on image processing 98 proceedings, Focus Interactive Technology Inc., Chicago, Illinois
3. Mansour M, Tewfik A (2002) Secure detection of public watermarks with fractal decision boundaries. In: Proceedings of the 11th European signal processing conference (EUSIPCO'02)
4. Choubassi ME, Moulin P (2005) A new sensitivity analysis attack. In: Proceedings of the SPIE international conference on security and watermarking of multimedia contents III. San Jose, CA, USA
5. Comesaña P, Pérez-Freire L, Pérez-González F (2005) The return of the sensitivity attack. In: Barni M (Hrsg.), Cox I (Hrsg.), Kalker T (Hrsg.), Kim HJ (Hrsg.) (eds) Digital watermarking: 4th international workshop, IWDW 2005 Bd. 3710, Lecture notes in computer science, Siena, Italy, Springer, pp 260–274
6. Comesaña P, Freire LP, Pérez-González F (2006) Blind newton sensitivity attack. IEEE Proc Inf Secur, 153(3):115–125
7. Kalker T (2001) Considerations on watermarking security. In: Proceedings of MMSP. Cannes, France, pp 201–206
8. Barni M, Bartolini F, Furon T (2003) A general framework for robust watermarking security. In: Signal processing, special issue on security of data hiding technologies
9. Diffie W, Hellman M (1976) New directions in cryptography. IEEE Trans Inf Theory, 22(6):644–654
10. Cox J, Miller M, Bloom J (2001) Digital watermarking. Morgan Kaufmann, Burlington
11. Pérez-Freire L, Pérez-Gonzalez F (2009) Spread spectrum watermarking security. IEEE Trans Inf Forensics Secur, 4(1):2–24
12. Cayre F, Fontaine C, Furon T (2005) Watermarking security: theory and practice. In: IEEE transactions on signal processing special issue "Supplement on Secure Media II"
13. Comesaña P, Pérez-Freire L, Pérez-González F (2005) Fundamentals of data hiding security and their application to Spread-Spectrum analysis. In: 7th information hiding workshop, IH05, Lecture notes in computer science. Springer, Barcelona, Spain

14. Pérez-Freire L, Pérez-González F (2008) Security of lattice-based data hiding against the watermarked only attack. IEEE Trans Inf Forensics Secur, 3(4):593–610. doi: 10.1109/TIFS.2008.2002938. ISSN 1556–6013
15. Pérez-Freire L, Pérez-González F, Comesaña P (2006) Secret dither estimation in lattice-quantization data hiding: a set-membership approach. In: Edward J, Delp III (Hrsg.) Wong PW (Hrsg.) (eds) Security, steganography, and watermarking of multimedia contents VIII. SPIE, San Jose, California
16. Kerckhoffs A (1883) La Cryptographie Militaire. Journal des Sciences Militaires Bd. IX:5–38
17. Doërr GJ, Dugelay, J-L (2004) Danger of low-dimensional watermarking subspaces. In: 29th IEEE international conference on acoustics, speech, and signal processing, ICASSP 2004, Montreal, Canada
18. Cachin C (1998) An information-theoretic model for steganography. In: Information hiding: second international workshop IHW'98. Portland, Oregon
19. Cayre F, Bas P (2008) Kerckhoffs-based embedding security classes for WOA data-hiding. IEEE Trans Inf Forensics Secur, 3(1)
20. Shannon CE (1949) Communication theory of secrecy systems. Bell Syst Tech J 28:656–715
21. Menezes AJ, Van Oorschot PC, Vanstone SA (1997) Handbook of Applied Cryptography. CRC, Boca Raton
22. Bogdanov A, Khovratovich D, Rechberger C (2011) Biclique cryptanalysis of the full AES. In: ASIACRYPT'11
23. Maurer U (2000) Authentication theory and hypothesis testing. IEEE Trans. Inf. Theory, 46(4):1350–1356
24. Cox I, Doerr G, Furon T (2006) Watermarking is not cryptography. In: Proceedings of the international work on digital watermarking, LNCS, Bd. 4283. Springer, Jeju island, Korea
25. Furon T, Bas P (2012) A new measure of watermarking security applied on DC-DM QIM. In: Information hiding 2012. Berkeley
26. Kay S (2000) Can detectability be improved by adding noise? IEEE Signal Process Lett 7(1):8–10. doi: 10.1109/97.809511, ISSN 1070–9908
27. Zozor S, Amblard P-O (2003) Stochastic resonance in locally optimal detectors. IEEE Trans Signal Process, 51(12):3177-3181. http://dx.doi.org/10.1109/TSP.2003.818905, doi: 10.1109/TSP.2003.818905
28. Cérou F, Del MP, Furon T, Guyader A (2011) Sequential Monte Carlo for rare event estimation. In: Statistics and computing, pp 1-14. http://dx.doi.org/10.1007/s11222-011-9231-6, doi: 10.1007/s11222-011-9231-6
29. Valizadeh A, Wang ZJ (2011) Correlation-and-bit-aware spread spectrum embedding for data hiding. In: IEEE Trans Inf Forensics Secur 6(2):267–282
30. Guyader A, Hengartner N, Matzner-Lober E (2011) Simulation and estimation of extreme quantiles and extreme probabilities. In: Appl Math Optim, 64:171-196. http://dx.doi.org/10.1007/s00245-011-9135-z, ISSN 0095–4616
31. Comesana P, Merhav N, Barni M (2010) Asymptotically optimum universal watermark embedding and detection in the high-SNR regime. IEEE Trans Inf Theory, 56(6):2804–2815
32. Furon T, Bas P (2008) Broken arrows. EURASIP J Inf Secur, 1–13. ISSN 1687–4161
33. Bas P (2011) Informed secure watermarking using optimal transport. In: IEEE proceedings of the international conference acoust, speech, signal processing
34. Furon T, Jégourel C, Guyader A, Cérou F (2009) Estimating the probability fo false alarm for a zero-bit watermarking technique. In: 2009 16th international conference on digital signal processing. IEEE, New York, pp 1–8
35. Bas P, Westfeld A (2009) Two key estimation techniques for the broken arrows watermarking scheme. In: Proceedings of the 11th ACM workshop on multimedia and security, MM and Sec '09. ACM, New York, pp 1–8. ISBN 978–1–60558–492–8
36. Bishop CM (1995) Neural networks for pattern recognition
37. Bas P, Furon T (2007) BOWS-2. http://bows2.ec-lille.fr

Chapter 4
Secure Design

This chapter provides different solutions to achieve a security level such as stego-security, subspace-security or key security while minimizing the embedding distortion.

Sections 4.1 and 4.2 present two solutions respectively for spread-spectrum and quantization embedding on synthetic host signals which are either Gaussian or Uniform, Sect. 4.3 presents the design of a watermarking scheme dedicated to original images which has been used for the 2nd international contest on Watermarking call BOWS (Break Our Watermarking System).

4.1 Secure Spread-Spectrum Embedding

4.1.1 Natural Watermarking

We design a secure SS-based watermarking scheme for the WOA framework. Natural Watermarking (NW) was named after its ability to (possibly) preserve the original pdf of the distribution of $z_{\mathbf{x},\mathbf{u}_i}$ during embedding, i.e. $z_{\mathbf{x},\mathbf{u}_i} \sim z_{\mathbf{y},\mathbf{u}_i}$. First, provided $\mathbf{x}$ has symmetrical pdf, one can easily show by Central Limit Theorem (CLT) argument that for N_v large enough:

$$z_{\mathbf{x},\mathbf{u}_i} \sim \mathcal{N}\left(0, \frac{\sigma_{\mathbf{x}}^2 \sigma_{\mathbf{u}_i}^2}{N_v}\right). \tag{4.1.1}$$

NW modulation uses side-information (SI) at the encoder to increase security, whereas it has been common during the last years to use it for increasing robustness.
NW modulation is defined as:

$$s_{NW}(\mathbf{m}[i]) = -\left(1 + \eta(-1)^{\mathbf{m}(i)} \frac{<\mathbf{x}|\mathbf{u}_i>}{|<\mathbf{x}|\mathbf{u}_i>|}\right) \frac{<\mathbf{x}|\mathbf{u}_i>}{\|\mathbf{u}_i\|^2}. \tag{4.1.2}$$

© The Author(s) 2016
P. Bas et al., *Watermarking Security*, SpringerBriefs in Signal Processing,
DOI 10.1007/978-981-10-0506-0_4

This modulation is more easily viewed as a model-based projection on the different vectors $\mathbf{u}_i$ followed by a η-scaling along the direction of $\mathbf{u}_i$. NW basically checks whether $z_{\mathbf{x},\mathbf{u}_i}$ lies on the desired side of the Gaussian curve. If not, it simply performs a model-based symmetry before applying a scaling. Also note that the condition for correct decoding is obviously:

$$\eta \geq 1. \tag{4.1.3}$$

From the security point of view, the original Bernouilli modulations are modified according to the values of the projection $z_{\mathbf{x},\mathbf{u}_i}$ which have Gaussian distribution. Again, by CLT argument, we have for N_v large enough:

$$s_{NW} \sim \mathcal{N}\left(0, \sigma_{s_{NW}}^2\right). \tag{4.1.4}$$

The fact that the sources in NW follow a Gaussian distribution ensures that NW is at least key-secure under our assumptions (WOA framework and independent messages), see Sect. 3.2. This clearly relates to the inability of ICA to separate sources in this case. Since we assume $\mathbf{x} \sim \mathcal{N}(0, \mathbf{I}_{N_v})$ ($\mathbf{I}_{N_v}$ is the identity matrix of size $N_v \times N_v$), one has obviously:

$$\mathbf{y}|\mathbf{K} \sim \mathcal{N}(0, \mathbf{J}_{N_v}),$$

with $\mathbf{J}_{N_v} = \mathbf{I}_{N_v}$ only if $\eta = 1$. This means that NW is stego-secure for $\eta = 1$. Otherwise it is just key-secure. Indeed, having $\eta > 1$ implies that subspace-security cannot be met.

We can show (see [1]) the following theoretical expectation of WCR for NW (which is actually a lower bound):

$$WCR_{NW} = 10 \log_{10}\left(\frac{(1+\eta^2)N_c}{N_v}\right). \tag{4.1.5}$$

Additionally, the expression of the BER for the AWGN channel is the following:

$$P_e = \int_0^{+\infty} \mathcal{G}_{\sigma_W^2 + \sigma_N^2}(t) \, \mathrm{erfc}\left(\frac{\sigma_W t}{\sqrt{2}\sigma_N\sqrt{\sigma_W^2 + \sigma_N^2}}\right) dt, \tag{4.1.6}$$

with $\mathcal{G}_{\sigma_W^2 + \sigma_N^2}(t) = \frac{1}{\sqrt{2\pi(\sigma_W^2 + \sigma_N^2)}} \exp\left(\frac{-t^2}{2(\sigma_W^2 + \sigma_N^2)}\right).$

If one wants to specify a target average WCR, the parameter η has the following expression:

$$\eta = \sqrt{\frac{N_v}{N_c} \times 10^{\frac{WCR}{10}} - 1}. \tag{4.1.7}$$

Therefore, the maximum number of bits to be securely hidden (i.e. $\eta = 1$) in **x** is:

$$N_c^{max} = \frac{N_v}{2} \times 10^{\frac{WCR}{10}}. \tag{4.1.8}$$

Figure 4.3 shows the performance of NW ($\eta = 1$) compared to other SS-based schemes. It is not surprising that SS always outperforms NW since it has a security constraint to meet that SS does not have to. Another remark is that NW does not achieve stego-security when the cover distribution is not Gaussian: in this case it only achieves subspace-security if $N_c = N_v$ or key-security if $N_c \neq N_v$. This means that for practical applications, NW can only be used to embed some hidden information into noisy components. When $\eta = 1$, NW simply amounts to the implementation of the well-known Householder reflection. We depict on Fig. 4.1 the distribution of the projection of two secret carriers on watermarked contents. We can see that neither cluster nor principal directions arise with NW, all other parameters being equal to SS and ISS embedding depicted in Fig. 5.2. It is not possible to estimate the secret keys because the Gaussian joint distribution of the projection of the carriers in the watermarked contents is circular (see below) and consequently any estimation of independent components (the carriers) is hopeless [2]. More importantly, circularity implies the definition of key-security since all the carriers that belong to the hypersphere provide the same density functions. Figure 4.1 presents the Gaussian joint distribution of two carriers. As we can see, it is not possible to find the directions that are associated to each carrier.

4.1.2 Circular Watermarking

Looking back at Figs. 5.2 and 4.1, it is clear that NW robustness can get much better by improving the separation of the decoding regions. It is the goal of the coding

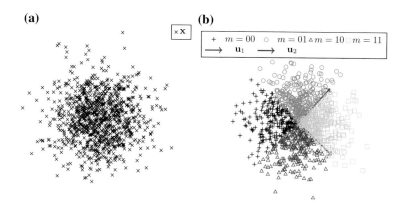

Fig. 4.1 Joint-distribution of two carriers for NW. NW does not produce any cluster, thus leading to stego-security. **a** Host distribution. **b** Distribution after natural watermarking

technique that we describe in this section under the term of Circular Watermarking (CW).

One can easily check that the joint-distribution of the projections of the secret carriers on the host signal using NW is circular.

Let $p(z_{\mathbf{x},\mathbf{u}_0}, \ldots, z_{\mathbf{x},\mathbf{u}_{N_c-1}})$ be the joint-distribution of the host on the carriers. Formally, we call *circular* any watermarking scheme which exhibits the following property:

$$p(z_{\mathbf{x},\mathbf{u}_0}, \ldots, z_{\mathbf{x},\mathbf{u}_{N_c-1}}) = p(\rho), \qquad (4.1.9)$$

where

$$\rho = \sqrt{\sum_{i=0}^{N_c-1} z_{\mathbf{x},\mathbf{u}_i}^2}.$$

Incidentally, note that NW is clearly circular because in that case we have $p(\rho) = \frac{1}{\sqrt{2\pi\sigma^2}^{N_c}} \exp\left(\frac{-\rho^2}{2\sigma^2}\right)$.

While Eq. (4.1.9) leaves many degrees of freedom for devising a circular watermarking scheme, we present here a practical implementation based on the well-known ISS modulation [3]. We could also have based our implementation on classical SS, but certainly at the cost of a lower robustness. For the sake of simplicity, we shall refer to this very implementation as CW in the sequel. The basic idea is to randomly spread the clusters of ISS (which are depicted on Fig. 5.2) on the whole decoding regions while preserving the circularity.

To this aim, let us construct a normalized vector $\mathbf{d} \in \mathbb{R}^{N_c}$ from another random vector $\mathbf{g} \sim \mathcal{N}(0, \mathbf{I}_{N_c})$. Each $\mathbf{d}(i)$ coefficient is constructed as follows:

$$\mathbf{d}[i] = \frac{|\mathbf{g}[i]|}{\|\mathbf{g}\|}. \qquad (4.1.10)$$

This vector will be independently drawn at each embedding and uniformly distributed on the positive orthant of the hypersphere. Our CW implementation requires exactly the same computations for ISS-parameters α and λ [3], which we intentionally omit here:

$$s_{CW}(\mathbf{m}[i]) = \alpha(-1)^{\mathbf{m}[i]}\mathbf{d}[i] - \lambda\frac{z_{\mathbf{x},\mathbf{u}_i}}{\|\mathbf{u}_i\|}. \qquad (4.1.11)$$

Naming of the vector $\mathbf{d}$ in Eq. (4.1.10) was chosen on purpose, since CW offers an analogy with the well-known DC-DM (Distorsion Compensated Dither Modulation, see Sect. 2.4.1) watermarking scheme where the dither is used to hide the location of the quantization cells. However, note that by construction CW is invariant to the scaling attack, contrarily to DC-DM schemes. We show on Fig. 4.2 the analogous of Figs. 5.2 and 4.1 for CW. Since Circular Watermarking renders carrier modulation jointly circular, we obtain dependency among message modulations, thus even powerful practical BSS attacks [4] are hopeless when using CW. There are no independent directions to allow for reliable carrier estimation using ICA.

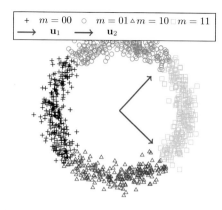

Fig. 4.2 Joint-distribution of the projection of two carriers for CW. The clusters of ISS have been spread over the entire corresponding decoding region thus leading to key-security

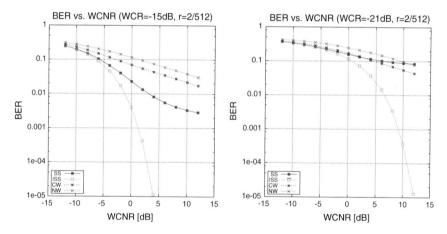

Fig. 4.3 BER comparison for SS, ISS, NW ($\eta = 1$) and CW

We depict on Fig. 4.3 the BER comparison between SS, ISS, NW and CW. We believe this figure points out what is the cost of true security for SS-based watermarking techniques. Interestingly enough, at typical WCR of -21 dB, CW performs close to SS but is always outperformed by ISS. At higher WCR however (-15 dB), the performance of CW degrades compared to other modulations.

4.1.3 Distribution Matching and Distortion Minimization

We present here a generalization of the secure embedding schemes that enables to fit a given distribution such as the one presented in Sect. 4.1.1 or 4.1.2. First we use a general secure embedding function, then we explain how it is possible to minimize the embedding distortion in this context.

First we assume that $\mathbf{x}$ is composed of samples $x[i]$ of variance σ_x^2, independent and identically distributed (iid) according to a distribution function $p_X(x)$.

The embedding of a message $\mathbf{m}$ is performed by applying an embedding function $f(.)$ on the host vector $\mathbf{x}$ to generate $\mathbf{y} = f(\mathbf{x}, \mathbf{m}, K)$ (K denoting the secret key). The watermark signal $\mathbf{w}$ is given by $\mathbf{w} = \mathbf{y} - \mathbf{x}$ and the variance of its samples is denoted σ_w^2.

We propose to use secure embedding functions $f(\mathbf{x}, \mathbf{m}, K)$, such that for all secret keys K:

$$p_X = p_{Y|K}. \tag{4.1.12}$$

These classes of embedding functions enable to obtain *stego-security* (see Sect. 3.2). We assume that the data-hiding scheme embeds a binary message with equal probability. In order to achieve *stego-security*, the density functions for each bit have to satisfy:

$$p_X = p_{Y|K} = (p_{Y|K,m=0} + p_{Y|K,m=1})/2. \tag{4.1.13}$$

One way to fulfill constraint (4.1.12) is to choose a partitioning function $g : \mathbb{R}^d \to \{0; 1\}$ such that $\int_{\mathbb{R}} g(\mathbf{x})p_X(\mathbf{x}) = 1/2$. $p_{Y|K,m=0}$ and $p_{Y|K,m=1}$ are consequently given by:

$$p_{Y|K,m=0}(\mathbf{x}) = 2g(\mathbf{x})p_X(\mathbf{x}), \tag{4.1.14}$$

and

$$p_{Y|K,m=1}(\mathbf{x}) = 2(1 - g(\mathbf{x}))p_X(\mathbf{x}). \tag{4.1.15}$$

The embedding function can now be considered as a set of two mapping functions $f_0(\mathbf{x}, K)$ and $f_1(\mathbf{x}, K)$ respectively for $m = 0$ and $m = 1$, satisfying respectively (4.1.14) and (4.1.15). The message decoding is performed using the partitioning function g:

$$g(\mathbf{z}) = 1 \Rightarrow \hat{m} = 0; \ g(\mathbf{z}) = 0 \Rightarrow \hat{m} = 1. \tag{4.1.16}$$

One straightforward solution to find mappings respecting constraint (4.1.12) is to choose g and generate random variables of pdf $p_{Y|K,m=0}$ and $p_{Y|K,m=1}$ (similar implementations have been proposed in [5] in the steganography context). However this solution doesn't take into account the distortion constraint ($D = \sigma_w^2$). The problem consists in finding mappings respecting (4.1.12) and (4.1.13) while minimizing the average embedding distortion.

One solution consists in using optimal transportation, also called optimal coupling, which has been defined by Monge in 1781 and consists in finding the transport from one density function p_1 to another p_2 that minimizes a cost function $c(\mathbf{x}_1, \mathbf{x}_2)$ representing the average transport [6]. Literally it is equivalent to the Monge's optimal transportation problem:

$$\text{Minimize} \int_X c(\mathbf{x}, f(\mathbf{x})) dp_1(\mathbf{x}) \tag{4.1.17}$$

over all the mapping functions f such that their transport on p_1 equals p_2.

If we consider the data-hiding framework we have $p_1 = p_X$ and $p_2 = p_{Y|K,m}$. Moreover we want to minimize the embedding distortion usually formulated as a quadratic cost: $c(\mathbf{x}, \mathbf{y}) = ||\mathbf{x} - \mathbf{y}||^2$. We can afterwards use the results of transportation theory in order to find the embedding functions $f_0(\mathbf{x}, K)$ and $f_1(\mathbf{x}, K)$ as the optimal mappings.

Given F_X the cumulative distribution function (cdf) of X, $F_{Y|K,m}$ the cdf of $Y|K$, m and $F_{Y|K,m}^{-1}$ the quantile function, one result of optimal transport is that for semi-continuous cdf, the optimal mapping for the quadratic cost is

$$f_{opt}(x, m) = F_{Y|K,m}^{-1} \circ F_X(x). \tag{4.1.18}$$

The minimum distortion is given by

$$D_{\min} = \int_0^1 c(F_{Y|K,m}^{-1}(x), F_X^{-1}(x)) dx. \tag{4.1.19}$$

Note that for $d > 1$, closed-form solutions provided by optimal transportation exist for mappings such as projections, radial functions or densities of independent variables [7].

4.1.4 Informed Secure Embedding for Gaussian Hosts

We assume in the sequel that the host components are iid Gaussian and that $d = 1$. This hypothesis can be practically verified since x can be the result of a projection of a vector (the wavelet coefficients of an image for example [8]) on a pseudo random carrier generated using the secret key K.

In the sequel, we use (4.1.18) to compute the optimal mapping, based on $F_X(x) = 0.5(1 + \text{erf}(x/\sqrt{2\sigma_x^2}))$. $F_{Y|K,m}(x)$ is computed for different partitioning functions, and message decoding is performed using Eq. (4.1.16). Without loss of generality, we assume $\sigma_x^2 = 1$. Three partitioning functions and associated embedding schemes are presented.

4.1.4.1 p-NW Embedding

In this case the real axis is split into $2M$ decoding areas of same probabilitybreak $p = 1/2M$. This can be achieved using as partitioning function:

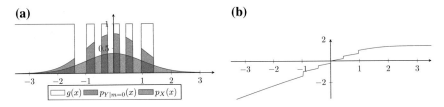

Fig. 4.4 p-NW for $M = 6$ and $\sigma_x^2 = 1$. **a** Partitioning function and associated pdfs. **b** Optimal mapping $f(x, 0)$

$$g(x) = 1 \quad \text{if } u(k) < x < u(k + 0.5),$$
$$g(x) = 0 \quad \text{if } u(k + 0.5) < x < u(k + 1).$$

$\qquad (4.1.20)$

where $u(k) = F_X^{-1}\left(\frac{k}{M}\right)$. The optimal mapping is then:

$$f_0(x) = F_X^{-1}\left(\frac{F_X(x)}{2} + \frac{\lfloor Mx \rfloor}{2M}\right),$$

$\qquad (4.1.21)$

$$f_1(x) = F_X^{-1}\left(\frac{F_X(x)}{2} + \frac{\lfloor Mx \rfloor + 1}{2M}\right).$$

$\qquad (4.1.22)$

where $\lfloor x \rfloor$ denotes the floor function.

Classical (non-informed) coding is equivalent to 2 decoding regions (one for 1, the other for 0) and $p = 0.5$. Transportation Natural Watermarking or TNW [9] becomes consequently a special case of p-NW.

The partitioning functions $g()$, p_X, $p_{Y|m=0}$ and the optimal mapping are depicted on Fig. 4.4.

4.1.4.2 $\bar{p}$-NW Embedding

This mapping can be seen has the symmetrised version of the previous one: the decoding regions are symmetrised according to the x-axis. This enables to have a decoding area around 0 which is twice the size of the decoding area for p-NW embedding. Now the partitioning function is:

$$g(x) = 1 \quad \text{if } u(k) < |x| < u(k + 0.5),$$
$$g(x) = 0 \quad \text{if } u(k + 0.5) < |x| < u(k + 1).$$

$\qquad (4.1.23)$

And the mapping consists here in using (4.1.21) if $x > 0$ (resp. $x < 0$) and $m = 0$ (resp. $m = 1$), or using (4.1.22) if $x > 0$ (resp. $x < 0$) and $m = 1$ (resp. $m = 0$).

4.1.4.3 Δ-NW Embedding

In order to transpose the Scalar Costa Scheme (SCS) [10] in the secure embedding context, we now choose g such that each decoded region has the same width Δ. Due to the symmetry of the Gaussian distribution, this is possible if

$$g(x) = \sum_k \Pi_\Delta \left(x - \frac{\Delta}{2} + 2k\Delta \right), \tag{4.1.24}$$

where $\Pi_\Delta(x)$ is the centered rectangular window function of width Δ. Using (4.1.18), the optimal mapping can then be expressed as:

$$f_0(x) = F_X^{-1} \left(\frac{F_X(x) - v_{2i}}{2} \right),$$

$$f_1(x) = F_X^{-1} \left(\frac{F_X(x) - w_{2i}}{2} \right). \tag{4.1.25}$$

Here $v_{2i} < F_X(x) < v_{2i} + 2F_X((2i+1)\Delta)$ for $m = 0$ and $w_{2i} < F_X(x) < w_{2i} + 2F_X(2i\Delta)$ for $m = 1$ with

$$v_{2i} = 2 \sum_{k=-\infty}^{i} [F_X((2k-1)\Delta) - F_X(2k\Delta)],$$

and

$$w_{2i} = 2 \sum_{k=-\infty}^{i} [F_X((2k-2)\Delta) - F_X((2k-1)\Delta)].$$

Note that contrary to p-NW and $\bar{p}$-NW where the embedding distortion cannot be chosen a-priori because they rely on the integer M, Δ-NW offers a continuous range of embedding distortions which are functions of the scalar Δ. The partitioning function and associated mapping for Δ-NW are illustrated in Fig. 4.5.

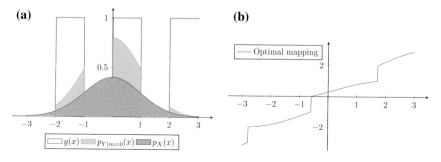

Fig. 4.5 **a** $g(x)$, $p_{Y|m=0}(x)$ and $p_X(x)$. **b** Optimal mapping for Δ-NW, $\Delta = 1$, $\sigma_x^2 = 1$, $m = 0$

4.1.4.4   Performance Comparison

We compare two different secure embedding schemes for the AWGN channel. In
order to provide a fair comparison, each scheme has to be evaluated for the same
embedding distortion. For comparison purposes, we also compute the Bit Error Rate
(*BER*) for two robust but insecure embedding schemes: Improved Spread Spectrum
(ISS) [3] and the Scalar Costa Scheme (SCS) [10].

Figure 4.6 compares the robustness of two implementations of Natural Watermarking: Transportation Natural Watermarking [9] uses optimal mapping but no
informed coding, in comparison Δ-NW uses optimal mapping and informed coding. For $WNR \geq -9$ dB the use of informed coding increases the robustness of the
scheme and the gap between the two implementations regarding the WNR is above
3 dB. Note that in this case the embedding distortion is fixed for TNW because it
depends only of σ_x^2.

Figures 4.7 and 4.8 present the performances of the different embeddings respectively for $WCR = -5$ dB and $WCR = -11$ dB. For p-NW and $\bar{p}$-NW these distortions are respectively equivalent to $M = 2$ and $M = 6$.

The performances of the secure embeddings differ according to the distortion but
general remarks can be drawn. For $WCR = -11$ dB and $WNR > -5$ dB, the Δ-NW

Fig. 4.6 *BER* of TNW [9]
and Δ-NW, $WCR = -0.7$ dB

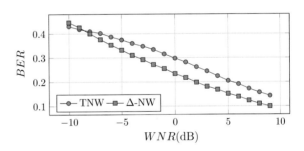

Fig. 4.7 Comparison
between secure and insecure
embeddings, $WCR = -5$ dB

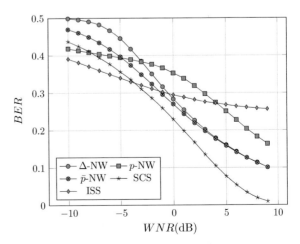

Fig. 4.8 Comparison
between secure and insecure
embeddings, $WCR = -11$ dB

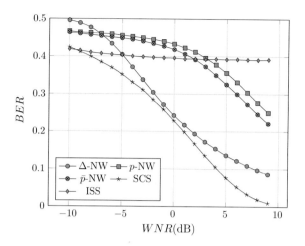

embedding is the secure scheme that provides the best performance and has a *BER*
close to SCS for $WNR = 0$ dB. Such behavior can be explained by the facts that for
$WNR = 0$ dB and low $WCRs$, (i) the distribution inside each decoding region can
be approximated as uniform and (ii) the embedding parameter for SCS is $\alpha = 0.52$
which is very close to the secure embedding parameter of SCS for uniform hosts
($\alpha_s = 0.5$) [11]. For such regimes SCS and Δ-NW embedding are consequently
very similar in term of robustness and security.

For $WCR = -5$ dB the decoding regions of Δ-NW are too far from each other and
the scheme provides poor performance in comparison with p-NW (for large WNR)
and $\bar{p}$-NW (for small WNR). Note also that for small WNR, the proposed secure-
embedding schemes can outperform the performance of the ISS (for both $WCRs$)
and SCS (for $WCR = -5$ dB).

4.2 Secure Quantization Based Embedding

The goal of this section is to design a new robust watermarking scheme for uniform
host secure under the WOA setup. Section 4.2.1 presents SCS, its robust implementations (e.g. enabling to maximize the transmission rate) and its secure implementations
(guarantying *perfect secrecy*). The maximum achievable rate for secure implementations is also analyzed for different Watermark to Noise Ratios ($WNRs$).

Section 4.2.2 proposes and extension of SCS presented in Sect. 2.4.2 called the
Soft-Scalar-Costa-Scheme (Soft-SCS) and the embedding and computation of the
distortion are detailed. Finally Sect. 4.2.4 compares the achievable rates of SCS and
Soft-SCS for both their secure and robust versions.

4.2.1   Scalar Costa Scheme

4.2.1.1   Notations

Here the subscript $._r$ denotes a *robust* implementation or parameter, e.g. the one maximizing the achievable rates and the subscript $._s$ denotes the *secure* implementation or parameter, e.g. satisfying the constraint of perfect secrecy. Hence SCS_r and SCS_s denote respectively robust and secure implementations of SCS which use respectively parameters α_r and α_s.

4.2.1.2   SCS Secure Modes

As it is mentioned in [12, 13], SCS achieves perfect secrecy under the WOA setup for an embedding parameter

$$\alpha_s = \frac{N_m - 1}{N_m}. \tag{4.2.1}$$

Indeed in this case we have $p_y = p_x$ and there is no information leakage about the location of the quantization cells. Additionally, the adversary is unable to distinguish watermarked samples from original ones. Two examples for $D = 2$ and $D = 3$ are illustrated on Fig. 4.9.

Equations (4.2.1) and (2.4.10) imply that one can maximize robustness while assuring perfect secrecy only if $\alpha_s = \alpha_r$, e.g. for a set of "secure" WNR_s equal to

$$WNR_s = -10 \log_{10} \left[\frac{1}{2.71} \left(\left(\frac{N_m}{N_m - 1} \right)^2 - 1 \right) \right]. \tag{4.2.2}$$

The range of WNR_s starts at -0.44 dB for $N_m = 2$ and $\alpha_s = 1/2$, consequently one way to perform both secure and robust watermarking is to select the alphabet

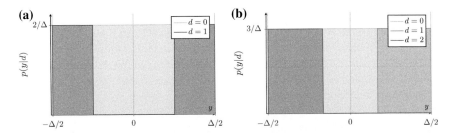

Fig. 4.9 Distributions of the watermarked contents for the two first secure modes of SCS. **a** $N_m = 2$, $\alpha = \frac{1}{2}$. **b** $N_m = 3$, $\alpha = \frac{2}{3}$

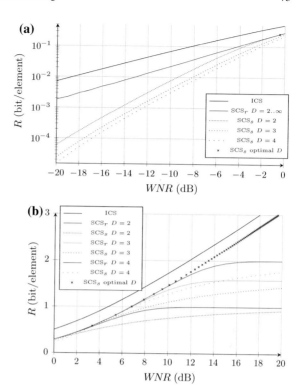

Fig. 4.10 Achievable rates for secure and robust SCS. The capacity of the Ideal Costa Scheme is also represented (here $D = N_m$). **a** Low WNR. **b** High WNR

size N_m which gives the closest WNR_s to the targeted WNR. However SCS doesn't offer efficient solutions for low WNR (e.g. < -1 dB).

In order to compare the performance of the secure version SCS_s and the robust version SCS_r we have computed the achievable rates using respectively α_r and α_s for a wide range of WNR and different alphabet size. The comparison is depicted on Fig. 4.10. All the rates are upper bounded by the Capacity of the Ideal Costa Scheme (ICS) $C_{ICS} = 0.5 \log_2(1 + 10^{WNR/10})$ [10, 14]. We can notice (Fig. 4.10a) that the performance gap between SCS_r and SCS_s is important for low WNR and it becomes negligible for high WNR (Fig. 4.10b), provided that the adequate alphabet size is selected. Note also that for a given N_m the gap between the secure and robust implementations grows with respect with the distance between the used WNR and WNR_s.

The inability of SCS_s to achieve efficient embedding for low WNR is due to the fact that SCS_r select a small embedding parameter α_r whereas SCS_s is lower bounded by $\alpha = 0.5$. The goal of the scheme presented in the next section is to modify SCS in such a way that the secure embedding provide better rates for low WNR.

4.2.2   Soft Scalar-Costa-Scheme

Contrary to classical watermarking embedding schemes, Soft-SCS is based on the principle of *optimal distribution matching*. In this context, the computation of the embedding can be seen as a two stage process. Firstly we setup the distribution $p_Y(y|d)$ of the watermarked contents, this first step is mandatory if one wants to create an embedding that achieves perfect secrecy. Secondly we compute the embedding that enables to match $p_Y(y|d)$ from the host signal of distribution $p_X(x)$ while minimizing the average distortion. This second step is performed using optimal transport theory (see Sect. 4.2.3).

Because the performances of SCS_s for low WNR are maximized for $N_m = 2$, the proposed scheme will be studied for binary embedding but could without loss of generality be extended to N_m-ary versions.

4.2.2.1   Shaping the Distributions of the Watermarked Contents

The rationale of Soft SCS is to mimic the behavior of SCS for $\alpha < 0.5$ while still granting the possibility to have perfect secrecy for given configurations. This is done by keeping the α parameter (we call it $\tilde{\alpha}$ in order to avoid confusion with the parameter used in SCS) and by adding a second parameter, called β, that will enable to have linear portions in the PDF of watermarked contents. β (respectively $-\beta$) are defined as the slope of the first (respectively the second) linear portions. The cases $\beta = +\infty$ is equivalent to SCS embedding. The differences between the distributions of watermarked contents for SCS and Soft-SCS are depicted on Fig. 4.11.

In order to fulfill the constraint that $\int_0^\Delta p_Y(y|m, y \in [0; \Delta])dy = 1$, the equation of the first affine portion on $[0; \Delta]$ is given by:

$$p_Y(y|m = 1, y \in [0; \Delta]) = \beta y + \frac{1 - \tilde{\alpha}(1 - \tilde{\alpha})\beta\Delta^2}{2(1 - \tilde{\alpha})\Delta} = \beta y + A, \qquad (4.2.3)$$

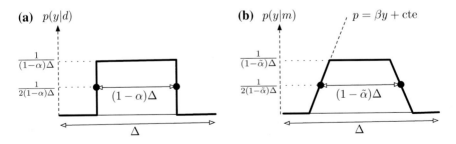

Fig. 4.11 Comparison between the distributions of **a** SCS and **b** Soft-SCS

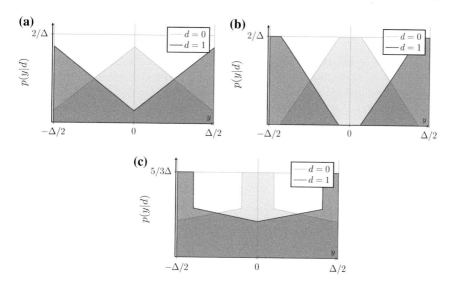

Fig. 4.12 Distributions of the watermarked contents for the 3 different configurations of Soft-SCS. **a** Big Top, $\tilde{\alpha} = \frac{1}{2}$, $\beta' = 0.4$. **b** Plateau, $\tilde{\alpha} = \frac{1}{2}$, $\beta' = 0.6$. **c** Canyon, $\tilde{\alpha} = \frac{2}{5}$, $\beta' = 0.1$

with $A = (1 - \tilde{\alpha}(1 - \tilde{\alpha})\beta\Delta^2)/(2(1 - \tilde{\alpha})\Delta)$ and by symmetry the second affine portion is gives $p_Y(y|m) = \beta(\Delta - y) + A$.

Depending of the values of $\tilde{\alpha}$ and β the distributions of $p_Y(y|m = 1, y \in [0; \Delta])$ for Soft-SCS can have three different shapes and the distributions will either look like a *big-top*, a *canyon* or a *plateau*. For illustration purpose, the 3 configurations are depicted on Fig. 4.12.

The intervals of the first linear portion (the second being computed by symmetry) and the type of shape are summarized in Table 4.1, they depend on a limit value of β called β_l which is different for $\tilde{\alpha} < 1/2$ or for $\tilde{\alpha} \geq 1/2$. For canyon and plateau shapes, the uniform portion of the PDF is equal to the one of SCS:

$$p_Y(y|m, y \in [0; \Delta]) = 1/((1 - \tilde{\alpha})\Delta). \tag{4.2.4}$$

Table 4.1 The different shapes of the distributions according to $\tilde{\alpha}$ and β

	$\tilde{\alpha} < 1/2$, $\beta_l = \frac{1}{\tilde{\alpha}(1-\tilde{\alpha})\Delta^2}$	$\tilde{\alpha} \geq 1/2$, $\beta_l = \frac{1}{(1-\tilde{\alpha}^2)\Delta^2}$
$\beta \leq \beta_l$	Canyon shape	Big Top shape
Domain of the affine portion	$[0; \tilde{\alpha}\Delta]$	$[(2\tilde{\alpha} - 1)\Delta/2; \Delta/2]$
$\beta > \beta_l$	Plateau shape	
Domain of the affine portion	$\left[\frac{\tilde{\alpha}\Delta}{2} - \frac{1}{2(1-\tilde{\alpha})\beta\Delta}; \frac{\tilde{\alpha}\Delta}{2} + \frac{1}{2(1-\tilde{\alpha})\beta\Delta} \right]$	

Note that only the Big Top shape and the Plateau shape can guaranty perfect secrecy

4.2.3 Embedding Computation and Decoding

The optimal way for computing the embedding that match the distribution of water-marked contents while minimizing the average distortion is to use the transportation theory [6, 9]. Given $F_Y(y|d)$ the CDF associated with $p_Y(y|d)$ and $F_X(x)$ the CDF associated with $p_X(x)$, the optimal transport minimizing the average L^2 distance is given by:

$$T(x) = F_Y^{-1} \circ F_X(x), \tag{4.2.5}$$

and the distortion by:

$$\sigma_w^2 = \int_0^1 (F_Y^{-1}(x|d) - F_X^{-1}(x))^2 dx. \tag{4.2.6}$$

Depending of the value of x, the transport is either non-linear:

$$T(x) = \frac{\nu_1 + \sqrt{\nu_2 + 2\beta(x - \nu_3)}}{\beta}, \tag{4.2.7}$$

or affine:

$$T(x) = (1 - \alpha)x + \frac{\alpha\Delta}{2}, \tag{4.2.8}$$

where ν_1, ν_2 and ν_3 are constants formulated in Table 4.2.

For visualization and parametrization purposes, since β ranges on $\mathbb{R}^+$ and depends on Δ, we prefer to use β' such that:

$$\beta = 4\tan\left(\pi\beta'/2\right)/\Delta^2, \tag{4.2.9}$$

where $\beta' \in [0, 1)$. The shape of the distribution becomes independent of Δ and the couple $\beta' = 0.5$ and $\tilde{\alpha} = 0.5$ corresponds to the case where the distribution $p_Y(y|d)$

Table 4.2 The different configurations for the computation of the distortion

	$\alpha < 1/2$	$\alpha \geq 1/2$
$\beta < \beta_l$	Canyon shape	Big Top shape
(x_0, x_1, x_2)	$(0;\ \beta\alpha^2\Delta^2/2 + \alpha\Delta A;\ 1/2)$	$(0;\ 1/2;\ 1/2)$
(ν_1, ν_2, ν_3)	$(-A, A^2, 0)$	$(-A, A^2, \nu_3)$
$\beta > \beta_l$	Plateau shape	Plateau shape
(x_0, x_1, x_2)	$(0;\ 1/\left(2(1-\alpha)^2\beta\Delta^2\right);\ 1/2)$	$(0;\ 1/\left(2(1-\alpha)^2\beta\Delta^2\right);\ 1/2)$
(ν_1, ν_2, ν_3)	$(-A, 0, 0)$	$(-A, 0, 0)$
β_l	$\frac{1}{\alpha(1-\alpha)\Delta^2}$	$\frac{1}{(1-\alpha^2)\Delta^2}$

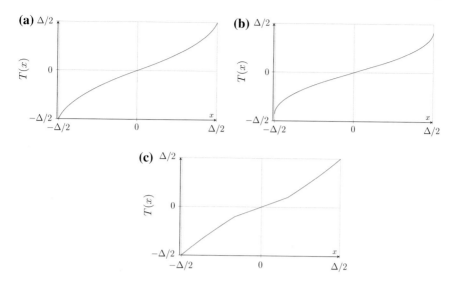

Fig. 4.13 Optimal transport for different configurations of Soft-SCS ($d = 0$). **a** $\tilde{\alpha} = \frac{1}{2}, \beta' = 0.4$. **b** $\tilde{\alpha} = \frac{1}{2}, \beta' = 0.6$. **c** $\tilde{\alpha} = \frac{2}{5}, \beta' = 0.1$

is at the junction between the big-top and the plateau. The cases $\beta' = 0$ and $\beta' \to 1$ correspond respectively to $\beta = 0$ and $\beta \to +\infty$.

Figure 4.13 illustrates different embeddings for $d = 0$ and different configurations of $(\tilde{\alpha}, \beta')$. Note that the embedding for $d \neq 0$ can be easily computed by translating both the host signal and the watermarked one by $\Delta/2$.

The embedding distortion is computed using Eq. (4.2.6) and contains 2 terms related respectively to the affine and non-linear portions of the embedding. Figure 4.14 illustrates the fit between the closed-form formulae and Monte-Carlo simulations.

As for SCS, the decoding is performed using maximum likelihood decoding.

4.2.4 Performance Analysis

4.2.4.1 Secure Embedding

It is easy to show that for $\tilde{\alpha} = \tilde{\alpha}_s = 0.5$ and $N_m = 2$, Soft-SCS achieves perfect secrecy, the distributions can only have two shapes in this case which are the *big-top* and the *plateau* illustrated on Fig. 4.12a, b respectively. Using numerical optimization, we compute for a given WNR the value of β' which enables to maximize the achievable rate (2.4.9) and obtain β'_s. The result of this optimization, and its approximation using least square regression is given on Fig. 4.15. The approximation gives

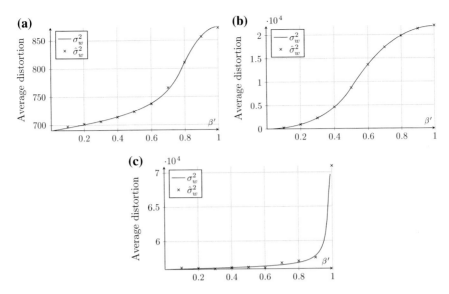

Fig. 4.14 Empirical distortions ($\hat{\sigma}_w^2$) computed by Monte-Carlo simulations with 10^6 trials, and closed-form distortions (σ_w^2) for $\Delta = 1024$, and 1024 bins used to compute the distributions. **a** $\tilde{\alpha} = 0.1$. **b** $\tilde{\alpha} = 0.5$. **c** $\tilde{\alpha} = 0.9$

Fig. 4.15 β_s' and its approximation

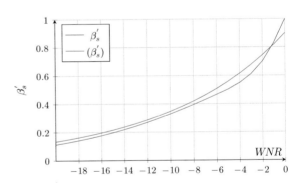

$$\begin{cases} (\beta_s') = 0.9 \times 1.1^{WNR}, & WNR < 0\,\text{dB} \\ (\beta_s') = 1, & WNR \geq 0\,\text{dB}. \end{cases} \tag{4.2.10}$$

which means that Soft-SCS$_s$ and SCS$_s$ differ only for $WNR < 0$ dB.

The achievable rates of Soft-SCS$_s$ are depicted on Fig. 4.16 and are compared with SCS$_r$ and SCS$_s$. We notice that Soft-SCS$_s$ not only outperforms the secure version of SCS but also the robust one. The gap between Soft-SCS$_s$ and SCS increases with respect to the noise power and is null for $WNR = -0.44$ dB. The figure shows also that the gap between the implementation for the optimal value of β_s' and the approximation given in (4.2.10) is negligible.

Fig. 4.16 Achievable rate of secure Soft-SCS

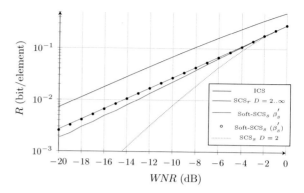

4.2.4.2 Robust Embedding

The same methodology is applied without the security constraint in order to obtain the robust configuration of Soft-SCS. This time the rate has to be maximized according to $\tilde{\alpha}$ and β' and their values after the numerical optimization are depicted on Fig. 4.17. For $WNR > -0$ dB, the values of β'_r oscillate between $\beta' = 0$ and $\beta' = 1$ which are two variations of SCS (the slope being null with a *big top* configuration or the slope being infinite with a *plateau* configuration).

We notice that there is no difference between Soft-SCS$_r$ and Soft-SCS$_s$ for $WNR < -9$ dB, the common optimal value being $\tilde{\alpha} = 0.5$ and the difference between the two schemes is negligible for $WNR < -0$ dB. For high WNR however, the approximation is identical to SCS$_r$ with $(\tilde{\alpha}_r) = \alpha_r$ (Eq. 2.4.10) and $(\beta'_r) = 1$. We can conclude that the implementation Soft-SCS$_r$ behaves as Soft-SCS$_s$ for low WNR and as SCS$_r$ for high WNR.

We have proposed in this section an adaptation of the Scalar Costa Scheme based on the principle of optimal distribution matching. The computation of the embedding needs (1) to choose the distribution of the watermarked contents and (2) to compute

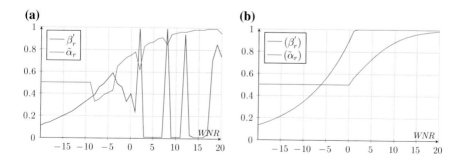

Fig. 4.17 Approximation of $\tilde{\alpha}_r$ and β'_t. **a** Optimal values for Soft-SCS$_r$. **b** Model

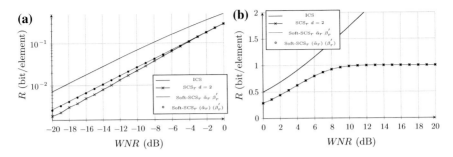

Fig. 4.18 Achievable rates for Soft-SCS$_r$. **a** Low *WNR*. **b** High *WNR*

the optimal mapping from the host to the watermarked contents. This method out-performs SCS both for its secure and robust implementations for $WNR \leq 0$ dB.

Contrary to a spread idea that robustness and security are antagonist constraints in watermarking, we have shown in this study that there exists watermarking schemes that can guaranty perfect secrecy while increasing the achievable rate. SCS$_s$ can be used for high *WNR* with appropriate dictionary size, $\alpha_s = (N_m - 1)/N_m$; and Soft-SCS$_s$ can be used for low *WNR*, $\tilde{\alpha}_s$ and β_s and provide negligible loss of rate (Fig. 4.18).

However, one can argue that for low *WNR* regimes the rates is rather small and that one system involving redundancy or error correction should be used in order to increase the reliability of the decoded symbols. This solution has to be employed in a very cautious way since the redundancy might compromise the security of the whole system [11].

4.3 Applied Watermarking Security

This section presents a practical watermarking scheme that was designed using several security principles such as a diversity in the generation of the key, an embedding built to resist to a Worst Case Attack and a detection region offering random boundary and local minima in order to deal with Oracle attacks.

The watermarking technique 'Broken Arrows' has been designed especially for the BOWS-2 (Break Our Watermarking Scheme 2nd Edition) contest. From the lessons learnt during BOWS-1, we had in mind to design a pure zero-bit water-marking scheme (no message decoding), which spreads the presence of the mark all over the host image. The BOWS-2 challenge is divided into three episodes with different contexts. The first episode aims at benchmarking the robustness of the technique against common image processing tools (compression, denoising, filtering...). Thus, 'BA' must be efficient so that it strongly multiplexes the original content and the watermarking signal in a non reversible way when the secret key is not known. Moreover, no robustness against geometrical attacks is needed because they yield low

PSNR values unacceptable in the contest. The second episode is dedicated to oracle attacks. The technique must be sufficiently simple so that the software implementation of the detector runs very fast because we expect a huge number of trials during this second episode. Counterattacks should be included if possible in the design. The third episode focuses on threats when many contents watermarked with the same secret key are released. The contenders are expected to deduce some knowledge about the secret key in order to better hack the pictures. 'BA' must not be trivially hacked. This is not an easy task especially since zero-bit watermarking tends to lack diversity.

In a nutshell, the detection regions are represented by a set of slightly modified hyper-cones. The embedding is classically done by moving a feature vector $\mathbf{x}$ of the host content deep inside this detection region to obtain a watermarked vector $\mathbf{y} = \mathbf{x} + \mathbf{w}$. The detection is performed by checking whether a feature vector $\mathbf{z}$ extracted from a submitted image belongs or not to one of these hyper-cones (see Sect. 4.3.2).

4.3.1 Four Nested Spaces

The embedding and the detection involve four nested spaces: the 'pixel' space, the 'wavelet' subspace, the 'correlation' subspace, and (what we call) the 'Miller, Cox and Bloom' plane (*abbr*. MCB plane). Index letters 'X, Y, W' denote respectively the representatives of the original content, the watermarked content, and the watermark signal to be embedded. We use the following terminology and notations to denote the representatives in the different domains:

- 'image' in the pixel space of width W_i and height H_i: $\mathbf{i}_Y, \mathbf{i}_X, \mathbf{i}_W$,
- 'signal' in the wavelet subspace, homomorphic to $\mathbb{R}^{N_s}$: $\mathbf{s}_Y, \mathbf{s}_X, \mathbf{s}_W$,
- 'vector' in the correlation space, homomorphic to $\mathbb{R}^{N_v}$: $\mathbf{y}, \mathbf{x}, \mathbf{w}$,
- 'coordinates' in the MCB plane, homomorphic to $\mathbb{R}^2$: $\mathbf{c}_Y, \mathbf{c}_X, \mathbf{c}_W$.

The diagram of the different processes necessary to obtain the different subspaces is depicted on Fig. 4.19. The following subsections describe these subspaces and their specificities.

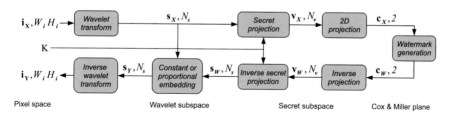

Fig. 4.19 The different processes and entities involved in the BA embedding scheme. Each couple $(\mathbf{a}, N_a)$ represents the name and size of the vector $\mathbf{a}$

4.3.1.1   The Pixel Space

Images are $H_i \times W_i$ matrices of 8-bit luminance values. We can always consider that the watermark in the pixel space is the difference between the watermarked image and the original image: $\mathbf{i}_W = \mathbf{i}_Y - \mathbf{i}_X$. This is not very useful, except that we impose a distortion constraint based on the PSNR, i.e. a logarithmic scale of the mean square error between pixel values of images $\mathbf{i}_X$ and $\mathbf{i}_Y$:

$$\text{PSNR} = 10 \log_{10} \frac{255^2}{\text{MSE}}, \tag{4.3.1}$$

where MSE is the mean square error: $\text{MSE} = (W_i H_i)^{-1} \sum_{i=1}^{H_i} \sum_{j=1}^{W_i} \mathbf{i}_W(i,j)^2$, with (W_i, H_i) being the width and height of the image in pixels.

4.3.1.2   The Wavelet Subspace

As stated in the introduction, the wavelet transform is an excellent embedding domain because of its compliance to human visual system.

We perform the 2D wavelet transform (Daubechies 9/7) on three levels of decomposition of the original image $\mathbf{i}_X$. This transform is very fast thanks to a very efficient lifted scheme implementation. We select the coefficients from all the bands except the low frequency LL band. These $N_s = W_i H_i (1 - 1/64)$ wavelet coefficients[1] are then stored into a signal $\mathbf{s}_X$ (a column vector). This signal lies in $\mathbb{R}^{N_s}$, a space we call the wavelet subspace. The low-low frequency band coefficients are kept in memory, and they will be used in the inverse extraction process.

The embedding process in this domain is in charge of mixing the host $\mathbf{s}_X$ and watermark $\mathbf{s}_W$ signals in a non reversible way. The result is the watermarked signal $\mathbf{s}_Y = f(\mathbf{s}_X, \mathbf{s}_W)$. We mean by non reversible the fact that an attacker observing $\mathbf{s}_Y$ should not be able to split it back to the two private signals.

The MSE in the wavelet subspace is equal to the MSE in the spatial domain because this wavelet transform conserves the Euclidean norm. Hence, to enforce the distortion constraint, we must have:

$$\|\mathbf{s}_Y - \mathbf{s}_X\| = 255 \sqrt{W_i H_i} \cdot 10^{-PSNR/20}. \tag{4.3.2}$$

4.3.1.3   The Secret Subspace

We use N_v secret binary antipodal carrier signals of size N_s: $\mathbf{s}_{C,j} \in \{-1/\sqrt{N_s}, 1/\sqrt{N_s}\}^{N_s}$. They are produced by a pseudo-random generator seeded by the secret key K. Their norm equals one, they are independent and we assume that they are orthogonal since their cross correlations are negligible (their expectations is zero,

---

[1] Image dimension must be multiple of 8.

and their standard deviations equal $1/\sqrt{N_s}$) compared to their unitary norms when N_s is important. The host signal is projected onto these carrier signals: $v_X(j) = \mathbf{s}_{C,j}^T \mathbf{s}_X$. These N_v correlations are stored into a vector $\mathbf{x} = (v_X(1), \ldots, v_X(N_v))^T$. It means that $\mathbf{x}$ represents the host signal in the secret subspace. We can write this projection with the $N_s \times N_v$ matrix $\mathbf{S}_C$ whose columns are the carrier signals: $\mathbf{x} = \mathbf{S}_C^T \mathbf{s}_X$. The norm is conserved because the secret carriers are assumed to constitute a basis of the secret subspace: $\|\mathbf{x}\|^2 = \mathbf{s}_X^T \mathbf{S}_C \mathbf{S}_C^T \mathbf{s}_X = \|\mathbf{s}_X\|^2$.

The secret subspace has several advantages. Its dimension is much lower than the wavelet subspace, the vectors in this space are easier to manipulate. It brings robustness against noise or, in other words, it increases the signal to noise power ratio at the detection side, because the noise is not coherently projected onto the secret subspace. Moreover, it boils down the strong non stationarity of the wavelet coefficients: components of $\mathbf{x}$ are almost independent and identically distributed as Gaussian random variables.

4.3.1.4 The Miller, Cox and Bloom plane

The MCB plane is the most convenient space because it enables a clear representation of the location of the hosts, the watermarked contents, and the detection boundary. This eases the explanation of the embedding and the detection processes. It is an adaptive subspace of the secret subspace whose dimension is two. We mean by adaptive the fact that this subspace strongly depends on the host vector $\mathbf{x}$. Denote $\mathbf{v}_c^\star \in \mathbb{R}^{N_v}$ a secret vector in the secret subspace, with unitary norm. A basis of the MCB plane is given by $(\mathbf{v}_1, \mathbf{v}_2)$ such that:

$$\mathbf{v}_1 = \mathbf{v}_c^\star, \quad \mathbf{v}_2 = \frac{\mathbf{x} - (\mathbf{x}^T \mathbf{v}_1)\mathbf{v}_1}{\|\mathbf{x} - (\mathbf{x}^T \mathbf{v}_1)\mathbf{v}_1\|}. \tag{4.3.3}$$

Hence, the MCB plane is the plane containing $\mathbf{v}_c^\star$ and $\mathbf{x}$. As far as we know, [15] is the first paper proposing the idea that the watermark vector should belong to the plane containing the secret and the host, hence the name MCB plane.

The coordinates representing the host is $\mathbf{c}_X = (c_X(1), c_X(2))^T$ with $c_X(1) = \mathbf{x}^T \mathbf{v}_1$, and $c_X(2) = \mathbf{x}^T \mathbf{v}_2$. Note that whereas $c_X(2)$ is always positive, the sign of $c_X(1)$ is not a priori fixed. However, we will define $\mathbf{v}_c^\star$ so that $c_X(1)$ is indeed always positive (see (4.3.17)).

An useful property of the MCB plane is that the norm of the host vector is conserved. The denominator of $\mathbf{v}_2$ can be written as:

$$\|\mathbf{x} - (\mathbf{x}^T \mathbf{v}_1)\mathbf{v}_1\|^2 = \|\mathbf{x}\|^2 + c_X(1)^2 - 2\mathbf{x}^T(\mathbf{x}^T \mathbf{v}_1)\mathbf{v}_1$$
$$= \|\mathbf{x}\|^2 - c_X(1)^2.$$

Hence,

$$c_X(2)^2 = \frac{(\|\mathbf{x}\|^2 - c_X(1)^2)^2}{\|\mathbf{x}\|^2 - c_X(1)^2} = \|\mathbf{x}\|^2 - c_X(1)^2, \qquad (4.3.4)$$

so that $\|\mathbf{c}_X\|^2 = c_X(1)^2 + c_X(2)^2 = \|\mathbf{x}\|^2$. Now, the vector $\mathbf{w}$ to be added in the secret subspace is indeed first generated in the MCB plane, such that $\mathbf{w} = c_W(1)\mathbf{v}_1 + c_W(2)\mathbf{v}_2$. Then, $\|\mathbf{w}\|^2 = \|\mathbf{c}_W\|^2$.

4.3.2 Embedding and Detection

As mentioned in the previous section, the embedding first needs to go from the spatial domain to the MCB plane. Then, it creates the watermark signal, and finally maps it back to the spatial domain. We have seen how to go from one subspace to another. We now explain the definition of the watermark representatives for the three domains.

We do not know what is the optimal way to watermark an image. This is mainly due to the non stationarity of this kind of host. However, as mentioned earlier, host vectors in the secret subspace can be modeled as random white Gaussian vectors. We know what is the optimal way (in some sense) to watermark a Gaussian white vector according to [16]. The embedder has to create a watermarked vector as $\mathbf{y} = a\mathbf{x} + b\mathbf{v}_c^\star$, where $\mathbf{v}_c^\star$ is a secret vector and a and b are scalars to be determined. This shows that the watermarked vector belongs to the plane $(\mathbf{x}, \mathbf{v}_c^\star)$, i.e. the MCB plane. However, contrary to [16], we prefer to look for the optimum watermarked coordinate in the basis $(\mathbf{v}_1, \mathbf{v}_2)$ of the MCB plane.

4.3.2.1 The MCB Plane

Knowing vector $\mathbf{v}_c^\star$, we perform the projection from the secret subspace to MCB plane as defined in (4.3.3). The detection region is defined by a cone of angle θ and abscissa direction $[0x)$ in the MCB plane such that $\mathbf{c}$ is considered watermarked if:

$$\frac{|(1,0) \cdot \mathbf{c}|}{\|\mathbf{c}\|} = \frac{|c(1)|}{\|\mathbf{c}\|} > \cos(\theta). \qquad (4.3.5)$$

The absolute value in the detection formula implies that the detection region is indeed a two-sheet cone as advised by N. Merhav and E. Sabbag [16].

The goal of the embedding process in this domain is to bring the coordinates of the watermarked vector $\mathbf{c}_Y$ deep inside the cone. There are actually several methods: maximizing a robustness criterion [17, Sect. 5.1.3], maximizing the detection score [16], or maximizing the error exponent [18]. These strategies assume different models of attack noise (resp. the noise vector is orthogonal to the MCB plane, the noise vector is null, or the noise vector is white and Gaussian distributed). We propose our own strategy which, in contrast, does not assume any model of attack as it foresees the worst possible noise. A geometrical interpretation makes the link between our strategy and the one from [17, Sect. 5.1.3].

4.3.2.2 Maximum Robustness

This strategy is detailed in [17, Sect. 5.1.3]. Assume that $c_X(1) > 0$. We look for an angle $\tau \in [0, -\pi/2]$ which pushes the watermarked coordinates deep inside the detection region. This operation is defined by:

$$\mathbf{c}_Y = \mathbf{c}_X + \mathbf{c}_W = \mathbf{c}_X + \rho(\cos(\tau), \sin(\tau))^T. \qquad (4.3.6)$$

The radius ρ is related to the embedding distortion constraint. We give its formula in Sect. 4.3.3.

Now, what does 'deep inside the cone' mean? Cox et al. propose to maximize a robustness criterion defined by:

$$R(\mathbf{c}_Y) = \max(0, c_Y(1)^2 \tan(\theta)^2 - c_Y(2)^2). \qquad (4.3.7)$$

Roughly speaking, R represent the amount of noise energy to go outside the detection region provided $\mathbf{c}_Y$ is inside [17, Sect. 5.1.3]. The maximum robustness strategy selects the angle $\tau^\star$ maximizing the robustness: $\tau^\star = \arg\max_{\tau \in [0, -\pi/2]} R(\mathbf{c}_Y)$, where $\mathbf{c}_Y$ is a function of τ (4.3.6). This can be done via a dichotomy search or a Newton algorithm.

We would like to give a geometrical interpretation of this robustness criterion. Assume first that the attack noise is independent of the secret vector $\mathbf{v}_c^\star$ and of the host vector $\mathbf{x}$. Geometrically speaking, it means that this noise vector $\mathbf{v}_N$ is orthogonal to the MCB plane, giving birth to orthogonal subspace spanned by $\mathbf{v}_3$ as depicted in Fig. 4.20a. A cut of the frontier of the detection region by the plane $(\mathbf{v}_2, \mathbf{v}_3)$ at the point $\mathbf{v}_Y$ shows a circle of radius $c_Y(1) \tan(\theta)$. Figure 4.20b shows that the square norm of $\mathbf{v}_N$ then needs to be at least equal to $(c_Y(1) \tan(\theta))^2 - c_Y(2)^2$ which is indeed $R(\mathbf{c}_Y)$.

4.3.2.3 A New Criterion Based on the Nearest Border Point Attack

The definition of the robustness explained above makes sense whenever the noise vector is orthogonal to the MCB plane. However, many attacks (filtering, compression,...) introduce a distortion which is indeed very dependent on the host vector. Hence, the previous assumption may not be realistic. We describe here a new embedding strategy maximizing the distance between the watermarked vector and the nearest border point on the detection region frontier. We first introduce it in an intuitive manner with geometrical arguments, and then we prove with a Lagrange resolution that this indeed is the best strategy.

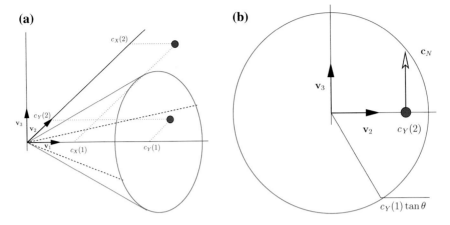

Fig. 4.20 Geometry in the MCB plane. **a** The hypercone and the MCB plane in the $(\mathbf{v}_1, \mathbf{v}_2, \mathbf{v}_3)$ basis. **b** The minimal norm attack vector $\mathbf{c}_N$, when it is orthogonal to $\mathbf{v}_1$ and $\mathbf{v}_2$

Fig. 4.21 The border point attack in the MCB plane

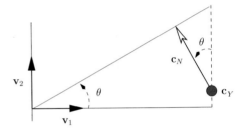

Assume that the noise vector belongs to the MCB plane, then the shortest path to move outside the detection region is to push the watermarked vector orthogonal to the edge of the cone as shown in Fig. 4.21. Hence, $\|\mathbf{c}_N\|^2 = (c_Y(1)\tan(\theta) - c_Y(2))^2 \cos(\theta)^2$. It is very easy to show that this norm is lower than $R(\mathbf{c}_Y)$. The embedding strategy should look for the coordinates $\mathbf{c}_Y$ maximizing the score $R'(\mathbf{c}_Y) = (c_Y(1)\tan(\theta) - c_Y(2))^2$ under the constraint that $\|\mathbf{c}_Y - \mathbf{c}_X\| = \rho$. In other words, we select the coordinates whose nearest border point attack needs the maximum noise energy. Intuitively, the embedder should push the host coordinates orthogonally to the edge of the cone so that: $\mathbf{c}_Y = \mathbf{c}_X + \rho(\sin\theta, -\cos\theta)^T$. However, this intuition is wrong when the embedding circle $\|\mathbf{c}_Y - \mathbf{c}_X\| = \rho$ intersects the axis of the cone because there is no point in having a negative $\mathbf{c}_Y(2)$ which would decrease $R'(\mathbf{c}_Y)$. This detail is illustrated in Fig. 4.22.

We now strengthen our rationale with a more rigorous approach. The noise vector can always be written as $\mathbf{v}_N = n_1\mathbf{v}_1 + n_2\mathbf{v}_2 + n_3\mathbf{v}_3$, where (n_1, n_2) are its coordinates in the MCB plane (the one defined by the original vector $\mathbf{v}_X$), and n_3 is the remaining component orthogonal to the MCB plane. Let us look for the nearest border point attack noise, that is finding which point $\mathbf{v}_N^{\star}$ located on the cone minimizes the Euclidean distance $\|\mathbf{v}_N - \mathbf{v}_Y\|$ to a point $\mathbf{v}_Y = (y_1, y_2, 0)$ in the MCB plane and inside the cone? This question can be mathematically formulated by:

$$(n_1, n_2, n_3)^\star = \arg \min_{n_1^2 \tan(\theta)^2 = n_2^2 + n_3^2} \|\mathbf{v}_N - \mathbf{v}_Y\|. \tag{4.3.8}$$

A Lagrange resolution gives a point $\mathbf{v}_N^\star$ function of the coordinates of $\mathbf{v}_Y$, and the minimum distance $d_{\min}(\mathbf{v}_Y) = \min(\|\mathbf{v}_N^\star - \mathbf{v}_Y\|)$. Keep in mind that our real job is to place in the MCB plane coordinate $\mathbf{c}_Y$ at a distance ρ from $\mathbf{c}_X$, while maximizing this minimal distance:

$$\mathbf{c}_Y = \arg \max_{(y_1,y_2):(y_1-c_X(1))^2+(y_2-c_X(2))^2=\rho^2} d_{\min}(\mathbf{v}_Y). \tag{4.3.9}$$

This second constrained optimization is also easily solved by a Lagrange resolution. The study is divided into two parts depending on the first Lagrange resolution:

Case (1) If $y_2 > 0$, the minimal distance equals $d_{\min}(\mathbf{v}_Y) = y_1 \sin(\theta) - y_2 \cos(\theta)$ which is positive since $\mathbf{v}_Y$ is inside the cone. The nearest border point belongs to the MCB plane (i.e. $n_3^\star = 0$) with coordinates $(n_1^\star, n_2^\star) = (y_1, y_2) + (-\tan(\theta), 1)(y_1 \tan(\theta) - y_2) \cos(\theta)^2$.

The second Lagrange resolution gives the watermarked coordinates: $\mathbf{c}_Y = \mathbf{c}_X + \rho(\sin(\theta), -\cos(\theta))$. Yet, this solution is acceptable only if $c_Y(2) > 0$, i.e. $c_X(2) > \rho\cos(\theta)$. The maximum of the minimum square distance is then $\max(d_{\min}^2) = (c_X(2) \cos\theta - c_X(1) \sin\theta - \rho)^2$. Vector instances are shown in Fig. 4.22 with superscript (1).

Case (2) If $y_2 = 0$ (i.e. $\mathbf{v}_Y$ is on the axis of the cone), then $d_{\min}(\mathbf{v}_Y) = y_1 \sin(\theta)$, and the nearest border points are located on a circle: $n_1^\star = y_1 \cos(\theta)^2$, and $n_2^{\star 2} + n_3^{\star 2} = n_1^{\star 2} \tan(\theta)^2$.

The distortion constraint allows to place the watermarked coordinates on the axis of the cone only if $c_X(2) < \rho$, and then $\max(d_{\min}^2) = (\sqrt{\rho^2 - c_X(2)^2} + c_X(1))^2 \sin(\theta)^2$. We re-discover here the embedding proposed in [16, Theorem 2], where optimal parameters are given by [16, (33)]. This is the 'erase' strategy where the embedder first erases the non coherent projection of the host and then spends the remaining distortion budget to emits a signal in the direction of the secret vector $\mathbf{v}_c^\star$. Vector instances are shown in Fig. 4.22 with superscript (2).

The two cases are possible and compete if $\rho\cos\theta \le c_X(2) \le \rho$. A development of the two expressions of $\max(d_{\min}^2)$ shows that, in this case, the first case gives the real maximum minimum distance. Denote $R' = \max d_{\min}^2$. Our embedder amounts to place $\mathbf{c}_Y$ to maximize this criterion:

If $c_X(2) \le \rho\cos\theta$, then

$$\mathbf{c}_Y = (c_X(1) + \sqrt{\rho^2 - c_X(2)^2}, 0)^T, \tag{4.3.10}$$

$$R' = (\sqrt{\rho^2 - c_X(2)^2} + c_X(1))^2 \sin(\theta)^2, \tag{4.3.11}$$

$$\mathbf{c}_N = c_Y(1) \sin(\theta)(-\sin(\theta), \cos(\theta))^T. \tag{4.3.12}$$

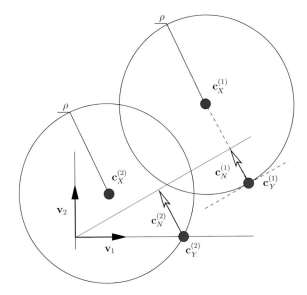

If $c_X(2) > \rho\cos(\theta)$, then

$$\mathbf{c}_Y = \mathbf{c}_X + \rho(\sin(\theta), -\cos(\theta))^T, \qquad (4.3.13)$$

$$R' = (c_X(1)\sin(\theta) - c_X(2)\cos(\theta) + \rho)^2, \qquad (4.3.14)$$

$$\mathbf{c}_N = (c_Y(1)\tan(\theta) - c_Y(2))\cos(\theta)^2(-\tan(\theta), 1)^T, \qquad (4.3.15)$$

where $\mathbf{c}_N$ is the the nearest border point attacked coordinate.

In conclusion, a geometrical interpretation of the robustness criterion R given in [17, Sect. 5.1.3] gives us an idea of changing it for R'. This idea has been checked via a double Lagrange optimization. This new formula has links with the embedding strategy of [16], and also avoids the iterative search, as the optimal watermarked coordinates have now closed form equation. The locus for the watermarked coordinates having the same robustness R' (aka, contour of constant robustness) is the cone translated by the vector $\sqrt{R'}/\sin\theta\,\mathbf{v}_1$. This is quite a different constant robustness surface that the hyperbola (defined through (4.3.7)), which gets very close to the border as the norm of the vector increases [15].

To go back to the wavelet subspace, we perform a double projection: firstly, we project $\mathbf{c}_W$ in the secret subspace and secondly, we project this result in the wavelet subspace to produce the watermark signal in the wavelet domain $\mathbf{s}_W$. Due to the use of orthonormal column vectors in both $\mathbf{S}_C$ (see Sect. 4.3.1.3) and $(\mathbf{v}_1, \mathbf{v}_2)$ matrices, this operation is defined by:

$$\mathbf{s}_W = \mathbf{S}_C(c_W(1)\mathbf{v}_1 + c_W(2)\mathbf{v}_2) = \mathbf{S}_C(\mathbf{v}_1, \mathbf{v}_2)\mathbf{c}_W. \qquad (4.3.16)$$

4.3.2.4 Increasing the Diversity of the Watermark Signal

Zero-bit watermarking is known to provide weak security levels due to its lack of diversity. The detection region, for the moment, is only composed of one two-nappe hypercone around the axis supported by $\mathbf{v}_c^\star$. Analyzing several watermarked signals, an attacker might disclose the secret signal $\mathbf{S}_C\mathbf{v}_c^\star$ that parametrized the detection region, using clustering or principal component analysis tools [4, 19, 20].

For this security reason, we render the detection region more complex, defining it as the union of several two-nappe hypercones. In the secret subspace, we define a set $\mathcal{C}$ of secret directions with N_c secret unitary vectors: $\mathcal{C} = \{\mathbf{v}_{C,k}\}_{k=1}^{N_c}$. The host signal being represented by $\mathbf{x}$ in this space, we look for the 'nearest' secret direction from the host vector:

$$\mathbf{v}_C^\star = \text{sign}(\mathbf{v}_C^T\mathbf{x})\mathbf{v}_C, \quad \text{with} \quad \mathbf{v}_C = \arg\max_{k \in \{1,\ldots,N_c\}} |\mathbf{x}^T\mathbf{v}_{C,k}|. \tag{4.3.17}$$

This secret vector is used for the embedding in the MCB plane. Chosen as is, projection $c_X(1)$ is always positive. At the detection side, the same vector has a high probability to be selected since the embedding increases correlation $\mathbf{y}^T\mathbf{v}_C'$.

We can predict two consequences. The first one is an advantage: We increase the probability of correct embedding. $\mathbf{v}_c^\star$ is chosen as the closest vector of $\mathcal{C}$ from $\mathbf{x}$, whence, for a given embedding distortion, it is more likely to push a vector $\mathbf{y}$ in its hypercone. This acceptance region split into several areas mimics the informed coding used in positive rate or zero rate watermarking scheme such as dirty paper trellis or quantized index modulation. The second one is a drawback: the angle of the cones decreases with the number of cones in order to maintain a probability of false alarm below a given significance level. Consequently, narrower hypercones yield a lower robustness, as less attack distortion is needed to go outside.

The following subsections investigate this issue from a theoretical and an experimental point of view.

4.3.2.5 Modeling the Host

In the wavelet subspace, one possible statistical model is to assume a Gaussian mixture: the wavelet coefficient $s_X(i)$ is Gaussian distributed but with its own variance σ_i^2: $s_X(i) \sim \mathcal{N}(0, \sigma_i^2)$. We will also pretend they are conditionally independent given their variances, which is of course not exactly true. In the secret subspace, the components of the host vector are Gaussian iid because the carrier signals are mutually independent, and the correlations are indeed linear combinations of Gaussian random variables. $v_X(j) \sim \mathcal{N}(0, \overline{\Sigma^2})$, with $\overline{\Sigma^2} = N_s^{-1}\sum_{i=1}^{N_s}\sigma_i^2$.

In the MCB plane, the statistical model is also very simple. Note first that $c_X(1)$ and $c_X(2)$ are not independent: $c_X(2) = \sqrt{c_X(1)^2 - \|\mathbf{x}\|^2}$. The first coordinate is defined as the maximum of N_c absolute values of correlation with unitary vectors. Hence, its cdf $F(c)$ is given by:

$$F(c) = \text{Prob}(c_X(1) < c) = \prod_{\mathbf{v} \in C} \text{Prob}(|\mathbf{v}^T \mathbf{x}| < c)$$

$$= \left(\Phi\left(\frac{c}{\sqrt{\overline{\Sigma^2}}} \right) - \Phi\left(\frac{-c}{\sqrt{\overline{\Sigma^2}}} \right) \right)^{N_c}$$

$$= \left(2\Phi\left(\frac{c}{\sqrt{\overline{\Sigma^2}}} \right) - 1 \right)^{N_c},$$

where Φ is the standard normal cdf. As N_c becomes larger, the Fisher-Tippet theorem shows that $F(c)$ converges to the Gumbel distribution with a variance decreasing to zero with a rate $1/\log N_c$. This allows us to roughly approximate $c_X(1)$ by its expectation which converges to the median value:

$$c_X(1) \sim \sqrt{\overline{\Sigma^2}} \Phi^{-1}((1/2^{1/N_c} + 1)/2), \qquad (4.3.18)$$

where Φ^{-1} is the inverse normal cdf. Approximating also $\|\mathbf{x}\|^2$ by $N_v \overline{\Sigma^2}$, (4.3.4) shows the following ratio is approximately constant:

$$c_X(2)/c_X(1) \sim \sqrt{\frac{N_v}{[\Phi^{-1}((1/2^{1/N_c} + 1)/2)]^2} - 1}. \qquad (4.3.19)$$

Hence, the locus of the host coordinates in the MCB plane focuses, more and more with N_c around a line passing by the origin and whose slope equals (4.3.19). Moreover, as mentioned above, the host coordinates get closer to the axis of the cone because the slope of the line is decreasing with N_c (see Fig. 4.23).

4.3.2.6   Probability of False Alarm

The hypercone is one of the very few detection regions where the probability of false alarm can be easily calculated provided that the host vectors pdf is radially symmetric, i.e. only depending of the norm of the vectors. This is the case in BA, we can thus use the work [21]: The probability that $\mathbf{x}$ falls inside a cone of angle θ is given by:

$$\text{Prob}(\mathbf{x} \text{ in a two-nappe cone}) = \frac{I_{N_v-2}(\theta)}{I_{N_v-2}(\pi/2)}, \qquad (4.3.20)$$

where $I_{N-2}(\theta)$ is the solid angle associated to angle θ in dimension N. We just bound P_{fa} with a classical union bound:

$$P_{fa} \leq N_c \frac{I_{N_v-2}(\theta)}{I_{N_v-2}(\pi/2)}. \tag{4.3.21}$$

As shown in Fig. 4.23, the angle of the cone is moderately decreasing with N_c.

4.3.2.7 Experimental Investigations

Figure 4.23 depicts the distributions of the coordinates of 5500 original images and their watermarked versions (PSNR of 45 dB) in *their* MCB plane for different numbers of cones while keeping the probability of false alarm below 10^{-6}. The host model is represented by the green line on the left. The bigger the number of cones the better the approximative model feats the experimental distribution. The effect of the proposed strategy to maximize the robustness is clearly visible: the points representing watermarked contents are either located on the $\mathbf{v}_1$ axis or nearly distributed along the blue line on the right parallel to the line modeling the host coordinates. Host coordinates above the dotted blue line are just shifted by the vector $\rho(\sin\theta, -\cos\theta)^T$.

Figure 4.24 represents the robustness criterion R' calculated for these 5500 real host coordinates against $\|\mathbf{c}_X\|$ and enables to draw important remarks for constant embedding:

- For images with a low magnitude of $\|\mathbf{c}_X\|$, the robustness decreases with the number of cones.
- For images with a high magnitude of $\|\mathbf{c}_X\|$, the robustness increases according to the number of cones.
- For a given number of cones, there is a range of $\|\mathbf{c}_X\|$, e.g. a class of images, where the robustness is maximal.
- The average robustness is monotonically increasing with the number of cones. It tends to saturate for $N_c > 50$. This is an extremely surprising experimental result

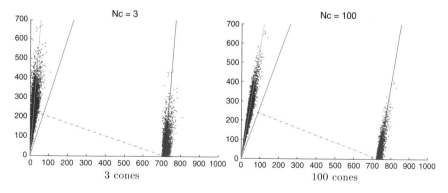

Fig. 4.23 Distribution of the hosts (*blue dots*) and their watermarked coordinates (*red dots*) in their respective MCB plane. $P_{fa} = 10^{-6}$, equivalent PSNR = 45 dB

Fig. 4.24 Computation of
the robustness R' in function
of $\|\mathbf{c}_X\|$ for different images
and different number of
cones. Average robustness
are represented by *horizontal
lines*

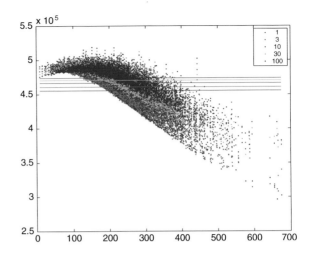

because we expected to have an optimal number of cones like the optimum number
of codewords in Costa's theory [14].

This subsection has investigated the watermark embedding and detection only in
the MCB plane. This exactly simulates an additive spread spectrum embedding where
the watermark signal defined in (4.3.16) is directly added to the wavelet coefficients:
$\mathbf{s}_Y = \mathbf{s}_X + \mathbf{s}_W$. This provides a tractable model in the MCB plane but it has many
drawbacks, as we shall see in the next section.

4.3.3   Proportional Embedding

The additive embedding has two major drawbacks. First, it doesn't comply with
some psychovisual basics. The power of the watermark signal is constant all over
the image, whereas the human eye is more sensitive on homogeneous regions than
on textured regions and edges. Experimentally, the watermark appears as noise over
uniform areas. The only way to avoid this artefact is to increase the PSNR, but then
the robustness becomes weak. A second drawback is that additive embedding does
not respect the Power Spectrum Condition [22] which states that the spectrum of
the watermark has to be locally proportional to the spectrum of the host in order
to be robust against denoising attacks. The intuition is that it is extremely hard or
almost impossible to filter out the watermark signal if it shares the same statistical
property than the host. A proportional embedding in the wavelet domain solves this
two issues. The proportional embedding consists in locally adapting a gain prior the
mixing: $\mathbf{s}_Y = \mathbf{s}_X + s_{W_p}$. The local gain is indeed proportional to the absolute value
of the host wavelet coefficient:

$$s_{W_p}(i) = |s_X(i)|s_W(i). \tag{4.3.22}$$

In other words, the signal s_W is hidden in the content via a proportional embedding. Such an embedding in the wavelet domain provides a simple Human Visual System in the sense that it yields perceptually acceptable watermarked pictures for PSNR above 40 dB [23]. Moreover, this scheme has shown to be close to the optimal embedding strategy given by a game theory approach, but less computationally expensive [24]. However, some corrections are needed in the BA algorithm.

4.3.3.1 Impact on Embedding Distortion

The following equation links the norms of $\|s_{W_p}\|$ and $\|s_W\|$, assuming that $s_X(i)$ is independent from $s_W(i)$, $\forall i \in [1, N_s]$:

$$\|s_{W_p}\|^2 = \sum_{i=1}^{N_s} |s_X(i)|^2 s_W(i)^2$$

$$\approx N_s^{-1} \sum_{i=1}^{N_s} s_X(i)^2 \sum_{i=1}^{N_s} s_W(i)^2$$

$$= \overline{S_X^2} \|s_W\|^2, \tag{4.3.23}$$

with $\overline{S_X^2} = N_s^{-1} \sum_{i=1}^{N_s} s_X(i)^2$. Hence, with (4.3.2), we must to set the norm of s_W to:

$$\|s_W\| = \frac{255\sqrt{W_i H_i}}{\sqrt{\overline{S_X^2}}} 10^{-\text{PSNR}/20}. \tag{4.3.24}$$

4.3.3.2 Equivalent Projection

A difficulty stems from the fact that the proportional embedding is not a linear process. Assume that the embedder calculates watermarked coordinates c_Y in the MCB plane and it mixes the corresponding watermark signal in the wavelet subspace with the proportional embedding. When the detector projects the watermarked signal back to the MCB plane, it does not find the same watermarked representative c_Y. The watermarked signal is projected back onto the secret subspace in y such that:

$$v_Y(k) = v_X(k) + \sum_{i=1}^{N_s} \sum_{j=1}^{N_v} |s_X(i)|v_W(j)s_{C,j}(i)s_{C,k}(i). \tag{4.3.25}$$

We assume that the host wavelet coefficient $S_X(i)$ is statistically independent of the ith secret carriers samples in order to simplify this last expression in provided that

$\mathbf{s}_{C,j}^T \mathbf{s}_{C,k} = \delta_{j,k}$:

$$v_Y(k) \approx v_X(k) + v_W(k)\overline{|S_X|}, \qquad (4.3.26)$$

with $\overline{|S_X|} = N_s^{-1} \sum_{i=1}^{N_s} |s_X(i)|$.

At the embedding side, we take into account this phenomenon right in the MCB plane. We model it by searching the best watermark coordinates with a vector $\mathbf{c}_{W_p}$, which reflects the coordinates of the vector $\mathbf{c}_W$ after proportional embedding in the MCB plane. But, the coordinates to be projected back to the secret subspace is indeed: $\mathbf{c}_W = \mathbf{c}_{W_p}/\overline{|S_X|}$.

These two corrections imply that even with a constant PSNR, the norm of $\mathbf{c}_{W_p}$ is different from a host image to another:

$$\|\mathbf{c}_{W_p}\| = \frac{\overline{|S_X|}}{\sqrt{\overline{S_X^2}}} 255 \sqrt{W_i H_i} \, 10^{-\text{ PSNR}/20}. \qquad (4.3.27)$$

In the MCB plane, the ratio $\overline{|S_X|}/\sqrt{\overline{S_X^2}}$ is the only difference between the additive and the proportional embedding methods.

4.3.4 Experimental Investigations

The final experimental work is a benchmark of three watermarking techniques. We used 2000 luminance images of size 512×512. These pictures represent natural and urban landscapes, people, or objects, taken with many different cameras from 2 to 5 millions of pixels.

The PSNR of the watermarked pictures is in average 42.7 dB. The visual distortion are invisible for almost all images. Figure 4.25 illustrates this with the reference image 'Lena'. A careful inspection shows some light ringing effects around the left part of the hat. However, there exists pictures where the embedding produces unacceptable distortion as shown in Fig. 4.25. We explain this as follows: The common factor of these images is that they are composed of uniform areas (e.g. the sky) or textures with very low dynamic (e.g. the trees), and they have very few strong contours (the street lamp and the statue of Fig. 4.25). Then, for a given distortion budget, the proportional embedding does not spread the watermark energy all over the image because most wavelet coefficients are small, but it focuses the watermark energy on the very few strong wavelet coefficients. For the purpose of the challenge, we did not care of it but this drawback has to be improved for a real watermarking technique.

Four watermarking techniques with different embedding strategies have been benchmarked:

(1) maximization of the robustness criterion R defined by (4.3.7) with a proportional embedding, (2) maximization of the error exponent as detailed in [18],

(a)

(b)

Fig. 4.25 **a** The reference image 'Lena' watermarked at PSNR = 42.6 dB. **b** One of few images where the embedding provides a poor quality despite a PSNR of 42.7 dB. Ringing effect are visible around the statue and the street lamp

(3) maximization of the new robustness criterion R' with a proportional embedding,
(4) maximization of the new robustness criterion R' with an additive embedding.

We apply a set of 40 attacks mainly composed of combinations of JPEG and JPEG 2000 compressions at different quality factors, low-pass filtering, wavelet sub-band erasure, and a simple denoising algorithm. This latter consists in thresholding wavelet coefficients of 16 shifted versions of the image, afterward the inverse wavelet transforms are shifted back and averaged.

Figure 4.26 reports the impact of 15 most significant attacks on the four techniques. The probability of detecting the watermark (i.e. number of good detection divided by 2000) is plotted with respect to the average PSNR of the attacked images. Because these classical attacks produce almost the same average PSNR, the four points for a given attack are almost vertically aligned. Yet, the impact on the probability of detection is interesting: despite that the additive embedding allows bigger radius embedding circle in the MCB plane, this technique is the weakest. This stresses the fact that mixing signals with different statistical structures as for constant additive embedding is partly reversible. This is an Achilles' heel that even classical attacks take benefit of. Our embedding strategy gives on average better results than the ones of Cox et al. (4.3.7) and of Comesaña et al. [18]. Yet, the improvement is really weak.

4.3.5 Counter-Attacks

In the BOWS-2 contest, the Broken Arrows algorithm has to face attacks linked to security. The first one is the oracle attack whose goal is to disclose the shape of

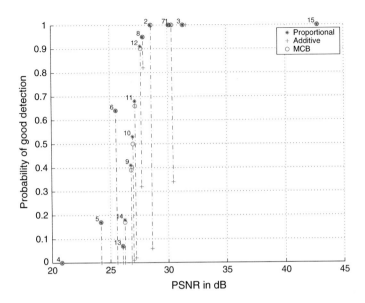

Fig. 4.26 Probability of good detection versus average PSNR of attacked images for the four water-marking techniques: proportional embedding and new robustness criterion '∗', additive embedding and new robustness criterion '+', proportional embedding and Miller, Cox and Bloom robustness criterion 'o' [17, Sect. 5.1.3], proportion embedding with Comesa?a's strategy 'Δ' [18]. Selection of attacks: *1* denoise threshold 20, *2* denoise threshold 30, *3* JPEG $Q = 20$, *4* JPEG2000 $r = 0.001$, *5* JPEG2000 $r = 0.003$, *6* JPEG2000 $r = 0.005$, *7* scale $1/2$, *8* scale $1/3$ + JPEG $Q = 50$, *9* scale $1/3$ + JPEG $Q = 50$, *10* scale $1/3$ + JPEG $Q = 60$, *11* scale $1/3$ + JPEG $Q = 70$, *12* scale $1/3$ + JPEG $Q = 90$, *13* scale $1/4$ + JPEG $Q = 70$, *14* scale $1/4$ + JPEG $Q = 80$, *15* no attack

the detection region and/or to find nearest border point. The second one is based on information leakage and the goal here is to try to estimate the secret subspace. We consequently decided to implement a counter-attack in order to make theses attacks (a bit) more complicated. The only solution we found[2] to cope with information leakage attacks is to increase the diversity of the key by using several cones (see Sect. 4.3.2.4). Regarding oracle attacks we adopted three counter-attacks presented below.

Randomized Boundary

An attacker having unlimited access to the detector as a black sealed box can lead oracle attacks. Many of them are based on the concept of sensitivity, where the attacker tries to disclose the tangent hyperplane locally around a point (called sensitive vector) on the border of the detection region. A counter-attack formalized by [25] is to slightly randomize the detection region for each call. This counter attack is very similar to the one that consists in having a chaotic boundary as proposed in

---

[2]Initially we also tried to increase the diversity of the key by using technique relying on perceptual hashing, but this technique was not mature enough to be implemented in the last final version of the algorithm.

[26, 27], both want to prevent an easy gradient ascent algorithm by making the detection border more difficult to analyze.

We process very simply by picking up a random threshold T uniformly distributed in the range $(\cos(\theta_{max}), \cos((\theta_{max} + \theta_{min})/2)]$. θ_{min} (resp. θ_{max}) has a corresponding probability of false alarm of 3×10^{-7} (resp. 3×10^{-6}).

Snake Traps

The 'snake' is a new kind of oracle attack invented by S. Craver et al. [28]. It consists in a random walk or a diffusion process in a constrained area of the space, which is indeed the detection region. This approach is a very efficient way to explore the detection region, and estimate parameters of the watermark detector. An important fact is that the snake tends to grow along the detection region border.

Our counterattack is to shape the boundary of the detection region, trapping the snakes in small regions to stop their growth. We draw 'teeth' in the MCB plane, in the following way:

- If $|c_Y(1) - \Delta\lfloor \frac{c_Y(1)}{\Delta} \rfloor| < r$, then detection is positive if $c_Y(1) > \|\mathbf{c}_Y\| \cos(\theta_{min})$,
- else the watermark is detected if $c_Y(1) > \|\mathbf{c}_Y\| \cos(\theta)$, where θ is a random variable as explained above.

Δ and r set the periodicity and the width of the 'teeth'. Note that the teeth are longer as the vector is far away from the origin. Depending on the step of the random walk, we hope to increase the probability of trapping a snake as it grows. The teeth slightly reduce the size of the acceptance region, hence, the probability of false alarm is even lower.

Snakes almost grow infinitely in a cone because this detection region is not bounded.[3] Hence, the average direction of several independent snakes can disclose the axis of the cone. Yet, we deal with several cones, and more importantly, the cones are indeed not disjoint for the considered probability of false alarm: the angle θ is always bigger than $\pi/4$ in Fig. 4.23 and around 1.2154 rad in the final implementation. The snakes will then be trapped in a subspace of dimension N_c, where no average direction will emerge. This doesn't mean however that snakes do no longer constitute a threat. A Principal Component Analysis of several long snakes might disclose the secret subspace. We expect at least a strong increase of detection trials.

Camouflage of the Cone

The detection score is virtually independent of a valuemetric scaling. This is a nice robustness feature, but very few detectors provide this advantage. Hence, this leaves clues [28]. We consequently decided to conceal the use of hypercones by truncating it: a content is deemed not watermarked if $\|\mathbf{c}_Y\| < \lambda$. Note that the value λ has to be small enough to guaranty that the nearest border point is not located on the truncated section of the cone.

These three countermeasures result in a detection region depicted in Fig. 4.27.

---

[3]In practice, the pixel luminance dynamic bounds it.

Fig. 4.27 Detection results in the MCB plane. *Dark-grey* points represent contents detected as watermarked, *light-grey* points as not watermarked. The angle of the cone has been chosen in order to magnify the shape of the border

Nasty Tricks

Concerning the challenge, we have the choice for the images proposed for the contest. We benchmark our watermarking technique over a set of 2000 images and against a bunch of common image processing attacks, in order to fine tune all the parameters, but also to investigate which images from this database were the most robust. These latter ones are used for the first episode of the challenge. In the same way, we made a light JPEG compression to let participants think that the embedding domain is the DCT domain!

The source code of the BA embedding and detection schemes and the script used to benchmark the algorithm are available on http://bows2.ec-lille.fr/BrokenArrowsV1_1.tgz.

The name 'Broken Arrows' comes from the fact that the detection region is a set of cones shaped like heads of arrows, where the very end has been broken (see Sect. 4.3.5).

Designing a practical watermarking technique for a contest is a very challenging task. We would like to point out that a design is necessary done under constraints of time, man and computer powers, with the sword of Damocles that a contender hacks the technique within the first hours of the challenge. Especially, the counter-measures presented in Sect. 4.3.5 have not been thoroughly tested due to lack of time. Consequently, a design is quite a different work than the writing of scientific paper. However, algorithms performing well in practice are often based on strong theoretical background.

Finally, now that the contest is over we can draw several conclusions:

1. Broken Arrows was rather robust, the best attack for the first episode achieved an average PSNR of 24.3 dB on the 3 images.
2. Broken Arrows was not secure against sensitivity attacks, the average PSNR achieved in this case was 50.24 dB, however the attack of the winners was rather time consuming to perform with around 4 millions calls to the detector.
3. Broken Arrows was not secure against information leakage attacks and the winner achieved an average PSNR of 46.21 dB. However the attack was not trivial to perform [29].

4.4 Conclusions of the Chapter

We must conclude this chapter by noting that the design of secure watermarking scheme is a complex process. If it is possible to achieve theoretical security on synthetic i.i.d. signals as proposed in the two first sections of this chapter, the distributions of real multimedia contents are usually more complex than i.i.d. Gaussian or i.i.d. Uniform and the security in this case is not proven anymore.

On the other hand, it is important to take into account security constraints when designing a watermarking scheme. As it is proposed in Sect. 4.3, the relatively weak solutions adopted to increase the security of a given watermarking scheme needed an important computational power in order to be broken [29] or were patched as in [30]. Even if not perfect, the relative security of a watermarking scheme is the first step toward the virtuous circle of attack and counter-attacks.

References

1. Cayre F, Bas P (2008) Kerckhoffs-based embedding security classes for WOA data-hiding. IEEE Trans Inf Forensics Secur 3(1):1–15
2. Hyvärinen A, Karhunen J, Oja E (2001) Independent component analysis. Wiley, New York
3. Malvar HS, Florencio DAF (2003) Improved spread spectrum: a new modulation technique for robust watermarking. IEEE Trans Signal Process 51:898–905
4. Cayre F, Fontaine C, Furon T (2005) Watermarking security: theory and practice. IEEE Trans Signal Process. Special issue "Supplement on secure media II"
5. Wang Y, Moulin P (2004) Steganalysis of block-structured stegotext. Security, steganography, and watermarking multimedia contents VI 5306:477–488
6. Villani C (2003) Topics in optimal transportation. American Mathematical Society, Providence
7. Cuesta-Albertos JA, Ruschendorf L, Tuerodiaz A (1993) Optimal coupling of multivariate distributions and stochastic processes. J Multivar Anal 46(2):335–361
8. Mathon B, Bas P, Cayre F, Macq B (2009) Comparison of secure spread-spectrum modulations applied to still image watermarking. Ann Telecommun 64(11–12):801–813
9. Mathon B, Bas P, Cayre F, Macq B (2009) Optimization of natural watermarking using transportation theory. In: MM&Sec'09: Proceedings of the 11th ACM workshop on multimedia and security. New York, NY, USA. ACM, pp 33–38 ISBN 978-1-60558-492-8
10. Eggers JJ, Buml R, Tzschoppe R, Girod B (2003) Scalar costa scheme for information embedding. IEEE Trans Signal Process 51(4):1003–1019
11. Pérez-Freire L, Pérez-González F, Furon T, Comesaña P (2006) Security of lattice-based data hiding against the known message attack. IEEE Trans. Inf Forensics Secur 1(4):421–439
12. Pérez-Freire L, Pérez-Gonzalez F (2009) Spread spectrum watermarking security. IEEE Trans Inf Forensics Secur 4(1):2–24
13. Guillon P, Furon T, Duhamel P (2002) Applied public-key steganography. In: Proceedings of SPIE: electronic imaging 2002, security and watermarking of multimedia contents IV Bd., vol 4675. San Jose, CA, USA
14. Costa M (1983) Writing on dirty paper. IEEE Trans Inf Theory 29(3):439–441
15. Miller ML, Doërr GJ, Cox IJ (2004) Applying informed coding and embedding to design a robust, high capacity watermark. IEEE Trans Image Process 6(13):791–807
16. Merhav N, Sabbag E (2008) Optimal watermarking embedding and detection strategies under limited detection resources. IEEE Trans Inf Theory 54(1):255–274
17. Cox J, Miller M, Bloom J (2001) Digital watermarking. Morgan Kaufmann, Amsterdam

18. Comesana P, Merhav N, Barni M (2010) Asymptotically optimum universal watermark embedding and detection in the high-SNR regime. IEEE Trans Inf Theory 56(6):2804–2815
19. Doërr GJ, Dugelay J-L (2004) Danger of low-dimensional watermarking subspaces. In: 29th IEEE international conference on acoustics, speech, and signal processing, ICASSP, 17–21 May 2004. Montreal, Canada
20. Bas P, Doërr G (2007) Practical security analysis of dirty paper trellis watermarking. In: Furon T, Cayre GDF, Cayre CF (Hrsg.), Bas P (Hrsg.) Information hiding: 9th international workshop. Saint-Malo, France. Springer, Berlin, Germany (Lecture notes in computer science)
21. Miller M, Bloom J (2000) Computing the probability of false watermark detection. Information hiding, Springer, New York
22. Su J, Girod B (1999) Power-spectrun condition for energy-efficient watermarking. In: IEEE international conference on image processing ICIP1999. Kobe, Japan
23. Bartolini F, Barni M, Cappellini V, Piva A (1998) Mask building for perceptually hiding frequency embedded watermarks. In: IEEE international conference on image processing 98 proceedings. Chicago, Illinois, USA. Focus Interactive Technology Inc
24. Pateux S, Le Guelvouit G (2003) Practical watermarking scheme based on wide spread spectrum and game theory. Signal Process Image Commun 18:283–296
25. Choubassi ME, Moulin P (2006) On the fundamental tradeoff between watermark detection performance and robustness against sensitivity analysis attacks. In: Delp EJ (Hrsg.), Wong PW (Hrsg.) Security, steganography, and watermarking of multimedia content VIII Bd., vol 6072(1I). San Jose, CA, USA (Proceedings of SPIE-IS&T Electronic Imaging)
26. Mansour M, Tewfik A (2002) Secure detection of public watermarks with fractal decision boundaries. In: Proceedings of the 11th European signal processing conference (EUSIPCO'02)
27. Linnartz J-P, van Dijk M (1998) Analysis of the sensitivity attack against electronic watermarks in images. In: International Information Hiding Workshop
28. Craver S, Yu J (2006) Reverse-engineering a detector with false alarm. In: Delp EJ (Hrsg.), Wong PW (Hrsg.) Security, steganography and watermarking of multimedia contents IX Bd., vol 6505(0C). San Jose, CA, USA (Proceedings of SPIE-IS&T Electronic Imaging)
29. Bas P, Westfeld A (2009) Two key estimation techniques for the broken arrows watermarking scheme. In: MM&Sec'09: proceedings of the 11th ACM workshop on multimedia and security, vol 1–8. New York, NY, USA. ACM. ISBN 978-1-60558-492-8
30. Xie F, Furon T, Fontaine C (2010) Better security levels for broken arrows. In: IS & T/SPIE electronic imaging international society for optics and photonics, pp 75410H–75410H

Chapter 5
Attacks

This chapter presents different ways to attack watermarking schemes and to estimate the secret key. One of the pioneering key estimation attack [1] consisted in estimating a set of secret carriers used in Spread Spectrum using Independent Component Analysis. The second section proposes to estimate more complex secret keys generated from dirty paper watermarking schemes using clustering methods. The third section proposes a security attack on Broken Arrows, the system used during the BOWS2 contest using a subspace estimation approach (see Table 5.1).

5.1 SS-Based Embedding Security

5.1.1 Attacks Using Independent Component Analysis

The watermark embedding for SS states that BPSK-based spread-spectrum watermarking can be seen as a noisy mixture of carriers following the model $\mathbf{Y} = \mathbf{X} + \mathbf{US}$. Noise is the original content and the mixture is parameterized by the modulation of the message. In this setup the problem of carriers estimation is what is commonly known as blind source separation (BSS). Given proper a priori knowledge, one typically wants to recover $\mathbf{S}$ (the *sources* in BSS theory) and possibly $\mathbf{U}$ (the *mixing matrix* in BSS theory). It is very insightful to notice that on one hand one advantage of BSS theory is that it makes no assumption on $\mathbf{U}$ the mixing matrix, but only on $\mathbf{S}$, the sources. On the other hand, other methods may use the fact that the columns of $\mathbf{U}$ are orthogonal to perform its estimation [2].

Given our fundamental hypothesis that the messages are drawn independently and that the carriers are scaled orthonormal, the projection of each carrier gives independent components:

© The Author(s) 2016
P. Bas et al., *Watermarking Security*, SpringerBriefs in Signal Processing,
DOI 10.1007/978-981-10-0506-0_5

Table 5.1 Different attacks presented in this chapter

Scheme	Security pitfall	Proposed attack
SS, ISS (see Sect. 2.2)	Independent carriers	ICA (see Sect. 5.1.1)
Broken arrows (see Sect. 2.3)	Variance modification in a subspace	Online PCA (see Sect. 5.1.2)
DPT (see Sect. 2.4.3)	Cluster around codewords	K-Means (see Sect. 5.2)

$$p(<\mathbf{y}|\mathbf{u}_1>,\ldots, <\mathbf{y}|\mathbf{u}_{N_c}>) = \prod_{i=1}^{N_c} p(<\mathbf{y}|\mathbf{u}_i>),$$

and our attacker shall therefore rely on ICA (Independent Component Analysis) to achieve his goal.

To assess the insecurity of a SS-based technique, we have decided to adopt the following methodology:

1. We generate N_o observations of watermarked contents and generate the matrix of observations $\mathbf{Y}$.
2. We whiten the observed signals using Principal Component Analysis. A reduction of dimension is therefore performed to reduce the searching time. If we consider that each host signal is generated from an *i.i.d.* process, the subspace containing the secret key will be included into a N_c-dimensional space of different variance [3]. We consequently select the subspace generated by eigenvectors corresponding to the N_c highest eigenvalues.
3. We run the FastICA algorithm [4] on this subspace to estimate the independent components and the independent basis vectors (e.g. the secret carriers).
4. We compute the normalized correlation c between each original and estimated carriers. A value of c close to 1 means that the estimation of the component is accurate. An estimation close to 0 means that the estimation is erroneous. For $N_c = 2$, we may evaluate the estimation accuracy by plotting a 2D constellation of points of coordinates (c_1, c_2). A successful estimation will then provide a point close to one of the four cardinal points $(0, 1)$, $(0, -1)$, $(1, 0)$, $(-1, 0)$.

We apply this ICA-based carrier estimation for both SS and ISS embedding. Figure 5.1 depicts the normalized correlation between the original and estimated carriers. We can notice that the estimations are globally more accurate for SS than for ISS with the BSS technique we used. In this case, this is mainly due to the fact that the variance of the watermarked signal after ISS is smaller than that after SS and consequently reduces the accuracy of the subspace estimation. Both SS and ISS were experimented at the same level of distortion: the Watermark-to-Content Ratio (WCR) was set to -21 dB.

Note however that ICA algorithms also have the fundamental limitations (due to seeking independent components) which encompass those defined in the previous section:

- it cannot recover the correct ordering of the mixing matrix columns;
- it outputs vectors that are only collinear to the mixing matrix columns.

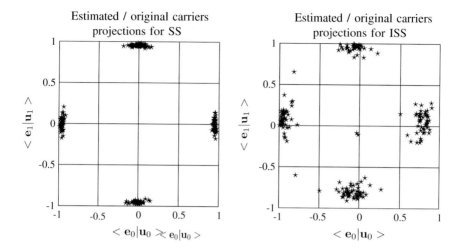

Fig. 5.1 Normalised correlations between the two estimated carriers and $\mathbf{u}_i \sim \mathcal{N}(0, 1)$ the real ones. For both schemes, $N_o = 1000$, $WCR = -21\,\mathrm{dB}$ and $N_v = 512$

This natural ambiguity means that in the WOA set-up, the set of carriers and their opposites will be considered as representing the very same key. In this context, Key-security (see Sect. 3.2.1.1) will be achieved only if it is impossible to estimate a secret carrier even up to a sign as shown in the following sections.

In Fig. 5.2, we depict the plot of $z_{\mathbf{y},\mathbf{u}_i}$ for traditional SS and ISS: when the sources are not Gaussian nor dependent, one can observe clusters oriented according to the

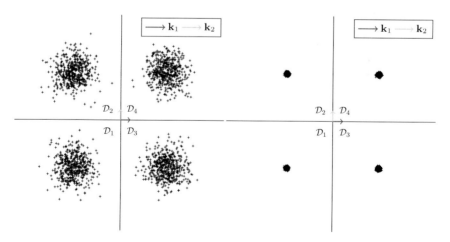

Fig. 5.2 Joint-distribution of two carriers projections for SS (*top*) and ISS (*bottom*). The formation of clusters allows to estimate the secret carriers by means of ICA. Watermarked contents projected on $\mathbf{k}_1$ and $\mathbf{k}_2$ ($N_b = 2$) and the associated decoding regions. $N_v = 128$, Gaussian host, DWR $= 5\,\mathrm{dB}$, $\gamma = 0.9$

positions of the secret carriers. Dependency and Gaussianity are two solutions to avoid such clusters to arise. Theses ideas are explained in the next section.

Toward secure embedding.

However it is important to point out that any ICA algorithm is known to fail (i.e. it outputs random sources and mixing matrix) in the two following cases:

- the sources are not independent,
- the sources are *i.i.d.* Gaussian signals of same variance.

Therefore, a successful approach to forbid accurate estimation of the carriers is to artificially make the sources become Gaussian and *i.i.d.* (Sect. 4.1.1: Natural Watermarking) or dependent (Sect. 4.1.2: Circular Watermarking).

5.1.2 *A Subspace Estimation Approach for Broken Arrows*

If the dimensionality of the secret subspace is too high, ICA is not effective enough and we have to use first a subspace estimation approach. This section illustrate this principle by presenting an approach based on a partial estimation of the secret projection used by the embedding algorithm Broken Arrows (see Sect. 4.3). Our rationale relies on the fact that the embedding increases considerably the variance of the contents within the secret subspace, in particular along the axes of the N_c vectors c_i that are used during the embedding. To illustrate this phenomenon, Fig. 5.3 depicts a comparison between the density of the maximum absolute correlations between c_i and original or watermarked contents on 10,000 images of the BOWS-2 challenge (embedding distortion of about 43 dB). This shows clearly an important increase of the variance within the secret subspace; consequently the strategy that is developed

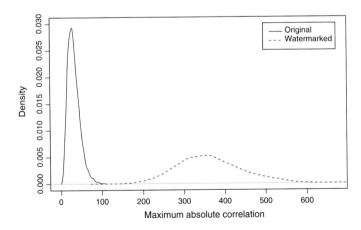

Fig. 5.3 Distribution of the maximum of the 30 correlations (in absolute value) for 10,000 images (PSNR = 43 dB), proportional embedding

in this section is to estimate the subspace spanned by the vectors $\{c_i\}$ by estimating the components of important variances from the observations.

If such similar strategies have already been used for security analysis of watermarking systems [1, 5, 6], the estimation of the secret subspace in our case is challenging for different reasons:

- Contrary to the systems addressed in [5, 6], the proposed method used a secret subspace of large dimension (30) in order to avoid basis estimation techniques such as averaging,
- The dimension of the host signal itself is very important (258,048),
- The system is used in real-life conditions on 10,000 images on a watermarking scheme that fulfills the different constraints regarding robustness but also visual distortion.

In order to perform subspace estimation, one usually uses Principal Component Analysis (PCA) which can be performed by an Eigen Decomposition (ED) of the covariance matrix obtained using the different observations. In our practical context however, the ED is difficult to perform because of the following computational considerations:

- The covariance matrix if of size $N_s \times N_s$, which means that 248 gigabytes are required if each element of the matrix is stored as a float,
- The computation of the covariance matrix requires around $O(N_o N_s^2) \approx 10^{12}$ flops,
- The computational cost of the ED is $O(N_s^3) \approx 10^{15}$ flops.

Consequently, we have looked for another way to compute the principal components of the space of watermarked contents. One interesting option is to use a inline algorithm which compute the principal vectors without computing any $N_s \times N_s$ matrices.

5.1.2.1 The OPAST Algorithm

The OPAST algorithm [7] (Orthogonal Projection Approximation Subspace Tracking) is a fast and efficient iterative algorithm that uses observations as inputs to extract the N_p principal components (e.g. the component associated with the N_p more important eigenvalues). The goal of the algorithm is to find the projection matrix $\mathbf{W}$ in order to minimize the projection error $J(\mathbf{W}) = E(||\mathbf{r} - \mathbf{W}\mathbf{W}^t\mathbf{r}||^2)$ on the estimated subspace for the set of observations $\{\mathbf{r}_i\}$. This algorithm can be decomposed into eight steps summed-up in Algorithm 5.1.

The notations are the following: the projection matrix $\mathbf{W}_0$ is $N_s \times N_p$ and is initiated randomly, the parameter $\alpha \in [0, 1]$ is a forgetting factor, $\mathbf{y}$, $\mathbf{q}$ are N_p long vectors, $\mathbf{r}$, $\mathbf{p}$ and $\mathbf{p}'$ are N_s long vectors, $\mathbf{W}$ is a $N_s \times N_p$ matrix, $\mathbf{Z}$ is a $N_p \times N_p$ matrix.

Step 6 is an iterative approximation of a the covariance matrix for the N_p principal dimensions. Steps 7 and 8 are the translations of the orthogonalisation process.

Since the complexity of OPAST is only $4N_s N_p + O(N_p^2) \approx 10^7$ flops per iteration, the use of the OPAST algorithm is possible in our context. Furthermore, it is easy to

---

Algorithme 5.1 The OPAST algorithm

1: Initialise $\mathbf{W}_0$ randomly
2: **for all** observations $\mathbf{r}_i$ **do**
3:    $\mathbf{y}_i = \mathbf{W}_{i-1}^t \mathbf{r}_i$
4:    $\mathbf{q}_i = \frac{1}{\alpha} \mathbf{Z}_{i-1} \mathbf{y}_i$
5:    $\gamma_i = \frac{1}{1+\mathbf{y}_i^t \mathbf{q}_i}$
6:    $\mathbf{p}_i = \gamma_i (\mathbf{r}_i - \mathbf{W}_{i-1}\mathbf{y}_i)$
7:    $\mathbf{Z}_i = \frac{1}{\alpha} \mathbf{Z}_{i-1} - \gamma_i \mathbf{q}_i \mathbf{q}_i^t$
8:    $\tau_i = \frac{1}{||\mathbf{q}_i||^2} \left(\frac{1}{\sqrt{1+||(p)_i||^2 ||\mathbf{q}_i||^2}} - 1 \right)$
9:    $\mathbf{p}_i' = \tau_i \mathbf{W}_{i-1}\mathbf{q}_i + \left(1 + \tau_i ||\mathbf{q}_i||^2\right) \mathbf{p}_i$
10:    $\mathbf{W}_i = \mathbf{W}_{i-1} + \mathbf{p}_i' \mathbf{q}_i^t$
11: **end for**

---

use and only relies on the parameter α for the approximation of the pseudo covariance matrix and does not suffer from instability.

5.1.2.2 Estimation Assessment

In order to run experiments and to assess the behavior of the subspace estimation algorithm we used the Square Chordal Distance (SCD) to compute a distance between two subspaces (the one coming from the secret key and the estimated subspace). The use of chordal distance for watermarking security analysis was first proposed by Pèrez-Freire et al. [8] and is convenient because SCD = 0 if the estimated subspaces are equal and SCD = N_p if they are orthogonal.

Given $\mathbf{C}$, a matrix with each column equal to one $\mathbf{c}_i$, the computation of the SCD is defined by the principal angles $[\theta_1 \ldots \theta_{N_c}]$ (the minimal angles between two orthogonal bases [9]) and their cosines are the singular values of $\mathbf{C}^t\mathbf{W}$ (note that this matrix is only $N_c \times N_p$):

$$\text{SCD} = \sum_{i=1}^{N_c} \sin^2(\theta_i) \tag{5.1.1}$$

A geometric illustration of the principal angles is depicted on Fig. 5.4.

Fig. 5.4 Principal angles between 2 planes π_1 and π_2

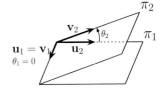

5.1.2.3 OPAST Applied on Broken Arrows

We present here the different issues that we have encountered and are specific to the studied embedding algorithm: the impact of the weighting method, the influence of the host signal and the possibility to use several times the same observations to refine the estimation of the secret subspace.

Constant Versus Proportional Embedding

We have first compared the impact of the embeddings given by the constant embedding and proportional embedding (Eq. 4.3.22). The behavior of the OPAST algorithm is radically different for these two strategies since the estimated subspace is very close to the secret subspace for constant embedding and nearly orthogonal to it for proportional embedding. The evolution of the SCD in both cases is depicted on Fig. 5.5.

Such a problematic behavior can be explained by the fact that the variance of the contents in the secret subspace is more important using constant embedding than using proportional embedding (compare Fig. 4.23 with Fig. 5.3).

The second explanation is the fact that the proportional embedding acts as a weighting mask which is different for each observation. This makes the principal directions less obvious to find since the added watermark is no more collinear to one secret projection.

Calibration

One solution to address this issue is to try to decrease the effect of the proportional weighting and to reduce the variance of the host signal. This can be done by feeding the OPAST algorithm with a calibrated observation $\hat{s}_Y$ where each sample is normalized by a prediction of the weighting factor $|s_X(i)|$ according to the neighborhood $\mathcal{N}$ of N samples:

$$\hat{s}_Y(i) = s_Y(i)/|\hat{s}_X(i)|, \tag{5.1.2}$$

Fig. 5.5 SCD for different embeddings and calibrations (PSNR $= 43$ dB, $N_p = 36$)

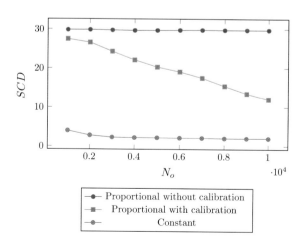

Proportional without calibration
Proportional with calibration
Constant

where

$$|\hat{s}_X(i)| = \frac{1}{N} \sum_{\mathcal{N}} |s_Y(i)|. \qquad (5.1.3)$$

The result of the calibration process on the estimation of the secret subspace is depicted on Fig. 5.5 using a 5×5 neighborhood for each subband. With calibration, the SCD decreases with the number of observations.

5.1.2.4 Principal Components Induced by the Subbands

Whenever watermarking is performed on non-iid signals like natural images, the key estimation process can face issues regarding interferences from the host-signals. Figure 5.6 depicts the cosine of the principal angles for $N_p = 30$ and $N_p = 36$ and one can see that all principal angles are small only for $N_p = 36$. For $N_p = 30$, only 25 out of 30 basis vectors of the subspace were accurately estimated. Consequently, depending on the embedding distortion, one might choose $N_p > N_c$.

In order to improve the estimation of the subspace, another option is to use the contents several times in order to obtain a better estimation of the pseudo-covariance matrix in the OPAST algorithm. Figure 5.7 shows the evolution of the SCD after

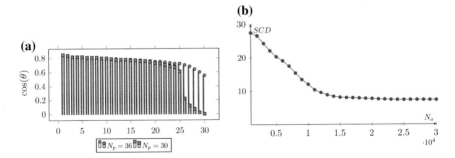

Fig. 5.6 Measuring subspace estimation. **a** $\cos \theta_i$ for $N_P = 36$ and $N_p = 30$, embedding PSNR $= 43$ dB. **b** Evolution of SCD after 3 runs (30,000 observations, PSNR $= 43$ dB, $N_p = 36$)

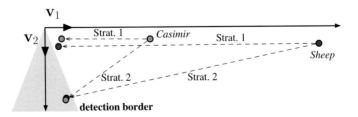

Fig. 5.7 Effects of the different strategies on the MCB plan for Casimir and Sheep

three multiple runs. We can notice that if the SCD decreases significantly between 10^4 and 2.10^4 observations, the gain for using a third run is poor though.

The last step of the key estimation process is to estimate each c_i by $\hat{c}_i$. Since all the variances along the different cone axes are equal, one solution to estimate the direction of each axis is to look for independent directions using Independent Component Analysis (see Sect. 5.1.1).

5.1.2.5 Leaving the Detection Region

The last step is to modify the watermarked content in order to push it outside the detection region of the hypercone of normalized axis $\hat{c}_k$ which is selected such that:

$$|s_Y^t \hat{c}_k| \geq_{j \neq k} |s_Y^t \hat{c}_i|.$$

Theoretically this is possible by canceling the projection between $\hat{c}_i$ and s_Y to create the attacked vector

$$s_Z = s_Y - \gamma s_Y^t \hat{c}_k \hat{c}_k.$$

However, practically $\hat{c}_k$ may not be accurate enough to be sure that $s_Z^t c_k = 0$, especially if the coordinates of the watermarked content are close to the cone axis. On Fig. 5.7, we can see the effect of this strategy (called "Strat. 1") on two images of the BOWS-2 contest Sheep and Casimir inside the MBC plan (the plan that includes c_k and the watermarked content s_Y). Another more efficient strategy is to push the content also to the directions that are orthogonal to $\hat{c}_k$; this can be done by increasing the projection of all the components except for the cone axis:

$$s_Z = s_Y - \gamma s_Y^t \hat{c}_k \hat{c}_k + \sum_{j \neq k} (\beta s_Y^t \hat{c}_j - 1)\hat{c}_j.$$

γ and β are constant factors specifying the amount of energy put in the directions which are respectively collinear and orthogonal to the cone axis. This second strategy (called Strat. 2) is depicted on Fig. 5.7 and the PSNR between the watermarked and attacked images for Sheep and Casimir are respectively equal to 41.83 and 48.96 dB.

Using the attack based on subspace estimation, the subspace is estimated on the 10,000 images provided by BOWS-2 contest. Each image is watermarked with a PSNR between 42.5 and 43 dB. As for Episode 3, proportional embedding is used.

OPAST is run using calibration on a 5×5 neighborhood for each subband, $N_p = 36$ and 2.10^4 observations (e.g. two runs), and the factor α is set to 1. The run of OPAST (two runs) takes approximately 6 hours on a 3 GHz Intel Xeon.

The ICA step was performed using fastICA [10], with a symmetric strategy and the tanh function to estimate negentropy. All the other parameters are set to defaults values.

Table 5.2 PSNR after successful attack using subspace estimation (i represents the number of iteration necessary to obtain a successful attack)

Image	PSNR (dB)	i	MCB coord.	MCB coord. after attack
Sheep	41.83	1	(925,48)	(62,223)
Bear	44.21	0	(532,47)	(88,253)
Horses	41.80	0	(915,20)	(77,233)
Louvre	48.95	0	(321,194)	(96,317)
Fall	46.76	0	(553,250)	(116,370)
Casimir	48.96	0	(352,31)	(59,234)

Watermark removal uses normalized estimated vectors $\hat{c}_i$ orientated such that $s_y^t \hat{c}_i > 0$. This strategy is used and the parameters are set to $\gamma = 1.1 + 0.1i$ (where i is a number of iterations) and $\beta = 50$.

The attack was performed on the five images used during the contest and available on the BOWS-2 website.

Table 5.2 presents the PSNRs after the attack and the number of necessary iterations. The coordinate of the original images in the MCB plane are also presented. As can be seen, the distortion is between 41.8 and 49 dB, which yields very small or imperceptible artifacts. Since the norm of the attacking depends of $s_y^t \hat{c}_i$, the farther the images are from the detection boundary, the more important the attacking distortion is.

5.2 Attacks on Dirty Paper Trellis Schemes

Attacking DPT watermarking schemes is more complicated than the previous schemes. First, the dimension of the subspace is higher since there are more codewords that possible messages, secondly contrary to SS schemes the watermark is highly dependent of the host content, finally we cannot observe a variance increase in a given subspace.

This security analysis assumes that some parameters of the DPT are public. It may not always be true in practice, but usually these parameters are fixed according to the desired robustness or payload of the algorithm. In this study for instance, the three parameters N_s, N_a and N_b will be known.

To define the secret key relative to a DPT watermarking scheme, it is necessary to identify which information is required by the attacker to perform security-oriented attacks such as:

- decoding the embedded message,
- altering the embedded message while producing the minimal possible distortion,
- copying the message to another content while producing the minimal possible distortion.

To decode the embedded message, Sect. 2.4.3 recalls that all parameters of the DPT are needed. This includes by definition

- the patterns attached to the arcs,
- the connectivity between the states,
- and the binary labels of the arcs.

To copy a message in an optimal way, it is first necessary to decode them and then to embed them into another content. Therefore, the same parameters are required. On the other hand, to alter the embedded message, all previous parameters are needed except the binary labels. Indeed, the watermarked vector $\mathbf{y}$ only need to be moved toward another vector $\mathbf{y}_A$ so that it no longer lies in the decoding region of $\mathbf{g}$. As long as a *neighbor* codeword is selected, it is unlikely to encode the same message and it will be close enough to avoid large distortion. This threat can be seen as the worst case attack (WCA) for DPT watermarking [11].

To perform this attack, it is necessary to know the two closest codewords from the watermarked vector $\mathbf{y}$, i.e. the embedded codeword $\mathbf{g}$ (see Sect. 2.4.3), and the second closest codeword from $\mathbf{y}$ ($\mathbf{b}_1$ in Fig. 5.8a). The attacker simply needs then to move the watermark content $\mathbf{y}$ somewhere inside the decoding region of this second best codeword ($\mathbf{y}_A$ in Fig. 5.8a) to make the detector fail while minimizing the distortion of the attack. In practice, this second best codeword is identified by feeding the Viterbi decoder with the watermarked vector $\mathbf{y}$ and successively forbidding a single step of the optimal path $\mathbf{g}$. This results in N_b candidates and the closest to $\mathbf{y}$ is retained as the second best codeword to be used in the WCA. This procedure is depicted in Fig. 5.8b.

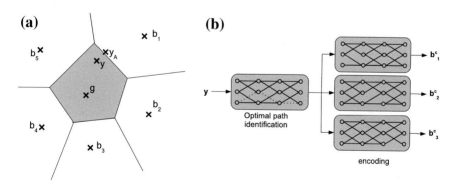

Fig. 5.8 Worst case attack for DPT watermarking. **a** To be optimal, the attacker needs to find the closest point to $\mathbf{y}$ outside the detection region of the embedded codeword $\mathbf{g}$ (*grey area*). To do so, he needs to identify the second nearest codeword from $\mathbf{y}$, i.e. $\mathbf{b}_1$ in the figure. **b** To identify the second best codeword, the Viterbi decoder is run several times with a single step of the optimal trellis forbidden. The codeword amongst the N_b candidates which is the closest to $\mathbf{y}$ is retained

5.2.1 DPT Parameters Estimation

For the sake of generality, the analysis will be done according to the Watermarked content Only Attack (WOA) setup [12], where the attacker observes different contents watermarked with different messages using the same secret key. The aim of this section is to present different techniques that can be used to estimate different parameters of a DPT that constitute the secret key, namely the patterns, the connectivity and the binary labels of the arcs.

Let $\mathcal{U} = \{\mathbf{u}_i, i \in [1, \ldots, N_a \cdot N_s]\}$ be the set of patterns, also referred to as carriers, associated with the arcs of the DPT. In practice, each pattern is usually normalised, e.g. $||\mathbf{u}_i|| = 1, \forall i$. As a result, each pattern can be seen as a point on the surface of the N_v-dimensional unit sphere. Moreover, each codeword $\mathbf{c}_i$ of the DPT is a $N_v \cdot N_b$-dimensional vector of norm $\sqrt{N_b}$ and can be considered as a point on the surface of a $N_v \cdot N_b$-dimensional sphere of radius $\sqrt{N_b}$, denoted $\mathcal{S}_c$.

Viterbi decoding aims at finding the codeword $\mathbf{c} \in \mathcal{C}$ which is the most correlated with some input vector $\mathbf{v}$, i.e. it evaluates and maximises:

$$\text{corr}(\mathbf{c}_i, \mathbf{v}) = \frac{< \mathbf{c}_i, \mathbf{v} >}{N_b \cdot N_v} = \frac{\sum_{j=1}^{N_b \cdot N_v} \mathbf{c}_i(j)\mathbf{v}(j)}{N_b \cdot N_v}, \tag{5.2.1}$$

which is equivalent to:

$$\text{corr}(\mathbf{c}_i, \mathbf{v}) = \frac{||\mathbf{v}||.||\mathbf{c}_i|| \cos(\theta_i)}{N_b \cdot N_v}, \tag{5.2.2}$$

where θ_i denotes the angle between $\mathbf{v}$ and $\mathbf{c}_i$. Because $||\mathbf{v}||$, $||\mathbf{c}_i||$ and $N_b \cdot N_v$ are constant terms, the codeword that is selected basically maximises $\cos(\theta_i)$.

In other words, the Viterbi decoder returns the codeword which is at the smallest angular distance from the input vector. This implies that when one wants to embed a codeword $\mathbf{g}$ in a host vector $\mathbf{x}$, it is necessary to produce a watermarked vector $\mathbf{y}$ whose angular distance with $\mathbf{g}$ is lower than with any other codeword in $\mathcal{C}$. Moreover, the higher the robustness constraint, the closer the watermarked contents to the desired codeword. Consequently, considering the distribution of normalized observations $\mathbf{y}^* = \mathbf{y}/||\mathbf{y}||$ one might observe *clusters* corresponding to the codewords in $\mathcal{C}$ on the surface of the $N_v \cdot N_b$ dimensional sphere $\mathcal{S}_c$.

5.2.2 Patterns Estimation Using a Clustering Method

Data clustering algorithms enable to analyse a large set of data by partitioning the set into subsets called clusters. Clusters are build such as to minimize the average distance between each data point and the nearest cluster center, also referred to as centroid. Given k the number of clusters, a clustering algorithm also returns the label of the centroid associated with each data point.

The K-means algorithm has been used here to provide a partition of the observed space. This algorithm labels each data point to the cluster whose centroid is the nearest. The centroid is defined as the center of mass of the points in the cluster, and its coordinates are given by the arithmetic mean of the coordinates of all the points in the cluster.

The implemented version of the algorithm was proposed by MacQueen [13] and is described below:

1. Choose k the number of clusters,
2. Initialise the centroids,
3. Assign each data point to the nearest centroid,
4. Update the centroid coordinates,
5. Go back to step 3 until some convergence criterion is met.

This algorithm is easy to implement, fast, and it is possible to run it on large datasets. However K-means does not yield the same result for each run, i.e. the final clusters depend on the initial random assignments. One solution to overcome this problem is to perform multiple runs with different initializations and to keep the result which provides the lowest intra-cluster variance. To ensure that the initial clusters are evenly distributed over the data set, a random initialization using the KZZ method [14] has been used.

5.2.2.1 Definition of the Dataset

A segment s is a portion of the observed watermarked vector y corresponding to a single step. Therefore, s is of size N_v and y is composed of N_b segments. Two alternative strategies are possible to estimate the secret parameters of the DPT:

(1) Apply the K-means algorithm to estimate the centroids representing the codewords of the trellis. Then it has to find $k = N_s \cdot N_a^{N_b}$ centroids in a $N_v \cdot N_b$-dimensional space using a dataset of normalised watermarked vectors.

(2) Apply the K-means algorithm to estimate the centroids representing the patterns of the trellis. Then it has to find $k = N_s \cdot N_a$ centroids in a N_v-dimensional space using a data-set of normalised watermarked segments.

Observing N_o watermarked contents is equivalent to observing $N_o \cdot N_b$ watermarked segments. As a result, the two strategies proposed earlier involve respectively $\frac{N_o}{N_s \cdot N_a^{N_b}}$ and $\frac{N_o \cdot N_b}{N_s \cdot N_a}$ observations per centroid. In other words, the second solution provides $N_b \cdot N_a^{N_b-1}$ times more observations per centroid than the first one to perform clustering. This problem is related to the *curse of dimensionality*, well known in machine learning, which states that the number of observations needed to learn topological objects such as clusters is exponential with respect to the dimension of the problem. Since the main concern here is the estimation of the patterns used in the DPT, the second solution is preferred to improve the estimation accuracy for the same number of observed contents.

5.2.2.2 Analysis of Estimation Accuracy According to Distortion

The accuracy of estimated patterns is inherently related to the embedding distortion, and therefore with the robustness constraint. Figure 5.9 depicts two typical examples of 3D distributions of normalized watermarked segments for two different embedding distortions. In this case only 6 codewords are used and one bit is embedded. The yellow balls indicate the centroids estimated using the K-means algorithm and the grey balls show the position of the true patterns. In this example, patterns are chosen to be either orthogonal or collinear (the set of orthogonal patterns is multiplied by -1 to obtain collinear ones). Each point of the distribution has a color depending on the center it has been associated.

In each case, the detection regions are represented by clusters which are clearly identifiable. Moreover the mapping between detection cells and embedded contents is consistent. However, for the smallest distortion (WCR = -6.6 dB), watermarked vectors are not uniformly distributed inside the embedding region. This is due to the fact that if two neighbor codewords encode the same message, their border region will have a density of codewords less important than if they encode different messages. This uneven distribution of watermarked codewords in each detection region results in a erroneous estimation of the codeword, the cluster center being "attracted" by the dense borders as illustrated on the right-hand distribution.

Figure 5.10 shows the accuracy of the DPT patterns estimation in the case of a realistic watermarking scenario. The different parameters of the trellis are defined here by $N_v = 12$, $N_b = 10$, $N_s = 6$, $N_a = 4$, which means that the clustering

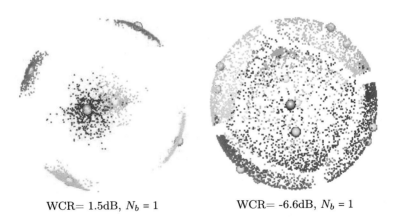

WCR= 1.5dB, $N_b = 1$ WCR= -6.6dB, $N_b = 1$

Fig. 5.9 Distributions of normalized watermarked contents ($N_v = 3$, $N_b = 1$, $N_s = 3$, $N_a = 2$, $N_o = 5000$). Locations of the real (*gray*) and estimated (*yellow*) patterns using the K-means algorithm

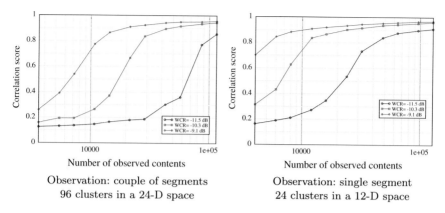

Fig. 5.10 Accuracy of the DPT patterns estimation ($N_v = 12, N_b = 10, N_s = 6, N_a = 4$). Average after 10 trials. For each trial, 10 K-means runs are performed

algorithm has to estimate 24 patterns of 12 samples each.[1] To evaluate the estimation process, the average of the difference between the two largest normalized correlations between each real and estimated patterns for each pattern is computed i.e.:

$$\Delta = \frac{1}{N_s \cdot N_a} \sum_i \left[\max 1_j(\text{corr}_N(\mathbf{cl}_i, \mathbf{u}_j)) - \max 2_j(\text{corr}_N(\mathbf{cl}_i, \mathbf{u}_j)) \right] \qquad (5.2.3)$$

where $\mathbf{cl}_i$ is the estimated centroid of the ith cluster, corr_N denotes the normalized correlation, and $\max 1_j$ (resp. $\max 2_j$) represents the first (resp. second) value when an array is sorted by descending order. As previously, the set of 24 patterns are orthogonal or collinear between themselves. As a result, Δ is equal to one if the estimation of each pattern is perfect and decreases with respect to the accuracy of estimations.

The difference between the two highest correlation scores magnifies the contrast between accurate and non-accurate estimation of the DPT patterns when they are orthogonal or opposite. A score Δ close to 1 means that each estimated pattern is much closer to one of the original patterns than the others. On the other hand, a score Δ close to 0 means that each estimated pattern is equally distant from two original patterns. Consequently, the estimation of the original patterns is not possible.

The evolution of the estimation accuracy with respect to different embedding distortions and different number of observations is given in Fig. 5.10 for observations composed of either one or two segments. If the considered dataset is composed of couples of segments, the number of observations necessary to obtain the same accuracy than for one segment is roughly multiplied by 4. This confirms the "curse of dimensionality" effect mentioned earlier. Moreover, as expected, the estimation accuracy increases with the number of observed contents and the embedding distortion

---

[1] $N_v = 12$ is the number of DCT coefficients that are used in the image watermarking scheme presented in [15].

i.e. the robustness constraint. While more than 12,8000 observations are needed to obtain an accuracy of 0.9 with $WCR = -11.5\,$dB and a data set of single segments, 24,000 and 8000 observations are needed respectively for $WCR = -10.3\,$dB and $WCR = -9.1\,$dB.

5.2.2.3   Connections and State Estimation

In the DPT estimation process, the next step is to estimate the connectivity of the trellis. This can be seen as learning which patterns are emitted at step $t + 1$ knowing that a given pattern has been emitted at step t. This estimation can be done by using a co-occurrence matrix $\mathbf{C}$ which is a square matrix of size $N_s \cdot N_a$. Each element $\mathbf{C}(i, j)$ of the matrix is expressed by:

$$\mathbf{C}(i,j) = \text{occ}(\mathbf{s}_t \in \mathcal{C}_i, \mathbf{s}_{t+1} \in \mathcal{C}_j) \tag{5.2.4}$$

where $\mathcal{C}_k$ denotes the set representing the kth cluster and $\text{occ}(A, B)$ is an occurrence function that counts the number of times both A and B are true. The test $(\mathbf{s}_t \in \mathcal{C}_i)$ is performed using the classification results of the K-means algorithm used for the patterns estimation. As a result, if the pattern i has been emitted at step t, the N_a maximum values in the ith row of the co-occurrence matrix $\mathbf{C}$ indicate the index of the patterns that can be emitted at step $t + 1$.

Using the established co-occurrence matrix, it is possible to check whether the estimated connectivity matches the one of the original trellis. For each line i in the matrix, the index of the N_a highest elements are retrieved. As stated before, each index points to the pattern that can be emitted at step $t + 1$ when the pattern i has been emitted at step t. This leads to $N_s \cdot N_a^2$ possible couple of patterns, that can be referred to as *connections*. The connection error rate is then defined as the ratio of connections which are actually not allowed by the original trellis. The lower the connection error rate, the more accurate the estimated connectivity. As depicted in Fig. 5.11, the accuracy relies again on the embedding distortion and the number of observed contents. It should be noted that the number of observed contents necessary to achieve a good estimation of the connections is of the same order of magnitude than for the estimation of patterns.

At this point, using the co-occurrence matrix, it is possible to identify for each pattern, which can also be viewed as an arc, which are the incoming and outgoing states. Each state is estimated up to a permutation with respect to the original trellis. However, this permutation does not hamper the ability of the decoder to retrieve the correct succession of patterns.

All the arcs going toward a given state will give similar rows in the co-occurrence matrix $\mathbf{C}$. Indeed, the rows indicate the choice of patterns that can be emitted afterward when an arc is traversed. Same rows implies same choice i.e. for all these arcs, the same state has been reached. To deal with the potential noise in the co-occurence matrix, a K-means algorithm is run on the rows of $\mathbf{C}$ to identify N_s clusters. Each row is then labeled in accordance to the cluster it belongs to. This label indicates

Fig. 5.11 Connection error rate ($N_v = 12$, $N_b = 10$, $N_s = 6$, $N_a = 4$). Observation: single segment. Average after 10 trials. For each trial, 10 K-means runs are performed

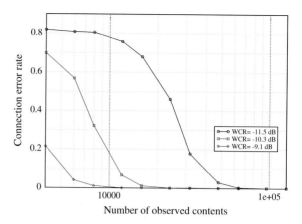

the outgoing state when an arc is traversed i.e. when a given pattern is emitted. For instance, in Fig. 5.12, if the third pattern is emitted, the systems reaches state 1 and can only emit the patterns 1 and 4.

One can then build an outgoing state matrix: it is a simple matrix with entries at the estimated connection index which indicates the outgoing state when the pattern i is emitted at step t. An example is given in Fig. 5.12. This matrix can be read row by row: if the pattern 3 is emitted at step t, then the system is in the third row and one can see that the state 1 is reached. Moreover, this outgoing state matrix can also be read column by column: if the pattern 3 has been emitted at step $t + 1$, then the system is in the third column and the entries indicates the possible states of the system before the emission of the pattern i.e. the incoming state. A simple majority vote along each column to accommodate for potentially noisy observations gives then the most likely incoming state for each pattern. In the example depicted in Fig. 5.12, one can see for instance that the pattern 3 is coming from state 2 and is going toward state 1.

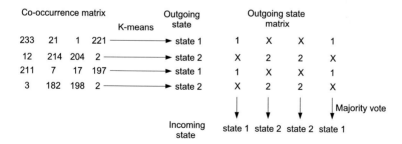

Fig. 5.12 Example of incoming and outgoing state estimations

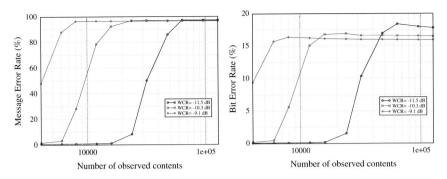

Fig. 5.13 Message error rate and Bit error rate after the WCA ($N_v = 12$, $N_b = 10$, $N_s = 6$, $N_a = 4$). Average after 10 trials. For each trial, 10 K-means are performed

5.2.2.4 Results on the Worst Case Attack

Once all the secret parameters of the DPT have been estimated, it is possible to perform the WCA. Results are plotted on Fig. 5.13 for the same setup than previously: $N_v = 12$, $N_b = 10$, $N_s = 6$, $N_a = 4$ and three different embedding distortions. Two different scores are computed to assess the efficiency of the WCA: the classical bit-error rate (BER) and the message error rate (MER). The MER is the most meaningful score because it measures the ability of the WCA to move the watermarked vector **y** outside the detection region of the embedded codeword. The BER plot in Fig. 5.13 highlights the fact that the WCA does not necessarily yield the same BER for different embedding distortions once the estimation of the trellis is accurate enough. Indeed, for different distortions, the second best codeword may be different and thus induce a different BER.

The security level s can be defined as the number of watermarked contents necessary to perform a successful WCA e.g. with a MER close to 100 %. The values of s for different embedding distortions are reported in Table 5.3. This table also provides a comparison of the average watermarked signal to noise ratios (SNR) for the WCA and AWGN required to yield equivalent BER. The WCA induces a SNR that is between 12 and 14 dB more important than for AWGN (the comparison between MERs would have been even more dramatic).

Table 5.3 Comparison of the security level and the signal to noise ratio of the WCA and AWGN for equal BER

Watermark to content ratio	−11.5 dB	−10.3 dB	−9.1 dB
Security level s	64×10^3	24×10^3	8×10^3
SNR for the WCA	16.9 dB	16.9 dB	16.9 dB
SNR for AWGN	4.5 dB	3.5 dB	2.4 dB

For SNR, an accurate estimation of the trellis ($N_o = 12,4000$) is performed

We can conclude this section by drawing several conclusions on the security of trellis based dirty paper coding:

- Using the WOA setup, it is impossible to estimate the binary labels associated with each arc of the trellis and consequently it is impossible to copy the message embedded in one content to another one without introducing unacceptable distortion. This property relies on the fact that coding is informed i.e. it is dependent of the host signal. Note that this property is not true for classical Spread Spectrum.
- The WOA setup enables however to perform a WCA for this scheme. Machine learning techniques can be used to identify clusters that are created in the data set during the embedding. This estimation has been performed using a K-means algorithm. Different tests suggest that an accurate estimation of the trellis is possible but depends on two parameters: the number of observations and the embedding distortion which is directly linked with the robustness of the scheme.

The assumptions made on the trellis structure may first look restrictive but encompass a large variety of practical implementations:

- The trellis structure was the same for each step. This hypothesis is important if one wants to deal with synchronization problems. Moreover, if it is not the case, because the trellis structure is the same for each content in the WOA setup, it is still possible to observe at least N_o similar segments (instead of $N_o \cdot N_b$) and to estimate the patterns for each step.
- The number of outgoing and incoming arcs per state was assumed to be constant. Nevertheless the presented connection and state estimation algorithms can also be used if the arcs change from one step to another.

5.3 Conclusions of the Chapter

This chapter shows that security analysis of watermarking schemes is linked with the problem of density estimation in machine learning and signal processing. In many watermarking schemes, because embedding regions have to be far from each other in order to guaranty robustness, watermarked contents are either gathered inside clusters in the media space $\mathcal{X}$ (this is the case for DPT watermarking) or are far from the distribution of the host content (this is the case for zero bit watermarking and SS). The adversary can either use clustering algorithms to estimate the secret key (see Sect. 5.2) or if the dimension of $\mathcal{X}$ is large the adversary has first to look for a subspace of smaller dimension in order to make the problem solvable (this strategy was used in Sects. 5.1.1 and 5.1.2). Of course, as shown in the previous chapter, secure embeddings are possible, but their robustness is in many cases less important.

References

1. Cayre F, Fontaine C, Furon T (2005) Watermarking security: theory and practice. In: IEEE transactions on signal processing, special issue supplement on secure media II
2. PËrez-Freire L, PËrez-Gonzalez F (2007) Disclosing secrets in watermarking and data-hiding. In: Third Wavila challenge, WACHA07. Saint-Malo, France, June 2007
3. Doërr GJ, Dugelay J-L (2004) Danger of low-dimensional watermarking subspaces. In: ICASSP 2004, 29th IEEE international conference on acoustics, speech, and signal processing. Montreal, Canada, 17–21 May 2004
4. Hyvärinen A (1999) Fast and robust fixed-point algorithms for independent component analysis. IEEE Trans Neural Netw 10(3):626–634
5. Doërr G, Dugelay J-L (2004) Security pitfalls of frame-by-frame approaches to video watermarking. IEEE Trans Signal Process 52(10):2955–2964
6. Cayre F, Bas P (2008) Kerckhoffs-based embedding security classes for WOA data-hiding. IEEE Trans Inf Forensics Secur 3(1):1–15
7. Abed-Meraim K, Chkeif A, Hua Y (2000) Fast orthogonal PAST algorithm. IEEE Signal Process Lett 7(3):60–62
8. Pérez-Freire L, Pérez-Gonzalez F (2009) Spread spectrum watermarking security. IEEE Trans Inf Forensics Secur 4(1):2–24
9. Knyazev AV, Argentati ME (2002) Principal angles between subspaces in an a-based scalar product. SIAM, J Sci Comput 23:2009–2041. Society for Industrial and Applied Mathematics
10. Hyvärinen A, Karhunen J, Oja E (2001) Independent component analysis. Wiley, New York
11. Vila JE, Voloshynovskiy S, Koval O, Perez-Gonzalez F, Pun T (2004) Worst case additive attack against quantization-based watermarking techniques. In: IEEE international workshop on multimedia signal processing (MMSP). Siena, Italy, 29Sep–1 Oct 2004
12. Cayre F, Fontaine C, Furon T (2005) Watermarking security part I: theory. In: Proceedings of SPIE, security, steganography and watermarking of multimedia contents VII, vol 5681. San Jose, USA
13. MacQueen J (1967) Some methods for classification and analysis of multivariate observations. In: LeCam LM, Neyman J (eds) Proceedings of the 5th Berkeley symposium on mathematics statistics and probability
14. He J, Lan M, Tan C-L, Sung S-Y, Low H-B (2004) Initialization of cluster refinement algorithms: a review and comparative study. In: Proceedings of IEEE international joint conference on neural networks, pp 25–29
15. Miller ML, Doërr GJ, Cox IJ (2004) Applying informed coding and embedding to design a Robust, high capacity watermark. IEEE Trans Image Process 13(6):791–807

Chapter 6
Conclusions and Open Problems

This chapter concludes this book on Watermarking Security and I hope that the reader will have a better view of the ins and outs of this domain. If watermarking security may look like a cat and mice game that is never ending, we can however state several important conclusions derived for the large variety of researches that have been conducted but there are also fascinating related problems that still need to be solved. The goal of this last chapter is to list the different outputs and open questions related to watermarking security but also general security of information forensics.

Conclusion 1 (connected with Chap. 3): there exits a methodology to analyze the security or increase the security of a watermarking scheme under a given scenario and given assumptions, this methodology takes into account the materials given to the adversary but also his goals and his power.

Conclusion 2 (connected with Chap. 3): several security proofs exist in watermarking security, such as the fact that an embedding is stego, subspace or key secure. These security proofs rely on non solved problems in signal processing, for example the fact that the separation of a mixture of Gaussian distributions of same variance is impossible, or the fact that if the host distribution is indistinguishable w.r.t. the watermarked one, the adversary can't estimate the secret key in the WOA setup.

Problem 1: most of the proofs rely on the fact that the host distribution is perfectly known which is hardly true in practice since this distribution is a high dimensional function which is difficult to infer. We have here the same limitation than in steganography, perfect security is only possible if the host distribution of the medium is completely defined. One consequence of this problem is that it is difficult to prove that a scheme is stego or subspace secure once we use real multimedia contents: the problems of watermarking security and density estimation are intertwined.

© The Author(s) 2016
P. Bas et al., *Watermarking Security*, SpringerBriefs in Signal Processing,
DOI 10.1007/978-981-10-0506-0_6

Conclusion 3 (connected with Chap. 4): if the security class is acknowledged, it is possible to minimize the embedding distortion using the results of optimal transport theory when the host distributions are simple (e.g. separable).

Problem 2: for non trivial distributions, optimal transport theory may not provide solutions. For example the problem of performing optimal transport for any circular distributions is as far as I know still open.

Conclusion 4 (connected with Chap. 4): it is possible to measure the security of a scheme using either the information leakage of the secret key, or the effective key length as a maximal bound.

Problem 3: how can we derive an optimal key estimator? Here the optimality means that the average number of trials to have access to the watermarking channel is minimum.

Problem 4: once the security (i.e. the difficulty to break the watermarking system) is measured, how to compute the "secure capacity" of a watermarking channel? i.e. the maximum amount of information that can be securely transmitted? Theoretical frameworks such as the wiretap channel theory may provide partial solutions in this case [1, 2].

Problem 5: this problem is linked with problem 4, how to perform a joint optimization that takes into account both distortion, capacity and security? Here, partial solutions might be found by looking at the connection with other theoretical tracks such as adversarial learning [3], adversarial signal processing [4], and game theory. For example, there are important similarities between adversarial learning [3] and watermarking secure to Oracle attacks [5, 6].

Problem 6: the computation of the effective key length is a first step of an approach for computational security in watermarking. How to compute a similar measure when the adversary generates his own observations at will and looks at the output of the decoder/detector, such as for an oracle attack? what is the effective key length in this case?

By looking at this list of open questions, we can see that the field of watermarking security need still important research efforts, and the conclusions illustrate the fact that more than a decade after its birth, watermarking is not, as suggested Herley a nonsense [7] but on the contrary makes more sense than ever [8] and has to be considered with all its specificities.

References

1. Wyner AD (1975) The wire-tap channel. Bell Syst Tech J 54(8):1355–1387
2. Ozarow LH, Wyner AD (1984) Wire-tap channel. II. AT & T Bell Lab Tech J 63(10):2135–2157
3. Lowd D, Meek C (2005) Adversarial learning. In: Proceedings of the eleventh ACM SIGKDD international conference on Knowledge discovery in data mining. ACM, pp 641–647
4. Barni M, Pérez-González F (2013) Coping with the enemy: advances in adversary-aware signal processing. In: Proceedings of IEEE international conferenceon acoustic, speech, and signal processing
5. Furon T, Bas P (2008) Broken arrows. EURASIP J Inf Sec 2008:1–13. ISSN 1687–4161

6. Linnartz JP, van Dijk M (1998) Analysis of the sensitivity attack against electronic watermarks in images. International information hiding workshop
7. Herley C (2002) Why watermarking is nonsense. Signal Process Mag IEEE 19(5):10–11
8. Moulin P (2003) Comments on "Why watermarking is nonsense". Signal Process Mag IEEE 20(6):57–59

Printed in the United States
By Bookmasters